Diseases of Vegetable Crops: Identification, Diagnosis and Treatment

Diseases of Vegetable Crops: Identification, Diagnosis and Treatment

Mahendra Kamalvanshi

RANDOM

RANDOM PUBLICATIONS
NEW DELHI (INDIA)

Diseases of Vegetable Crops: Identification, Diagnosis and Treatment

---

ISBN 978-93-5111-602-8

Published in 2015 in India by

RANDOM PUBLICATIONS

4376-A/4B, Gali Murari Lal, Ansari Road
New Delhi-110 002
Phone : +9111-43580356, 011-23289044, 011-43142548
e-mail: sales@randompublications.com,
info@randompublications.com, randomexports@gmail.com

Reprinted 2026

Type Setting by : Friends Media, Delhi-110089

Preface

Many plant diseases attack vegetables, but you can often avoid damage by careful husbandry or by applying control measures at appropriate times of the year. You can manage your vegetable garden to minimise losses if you understand the diseases and how to handle them.

Seeds of many vegetables are susceptible to damping-off fungi when planted in infested soils. The seeds may fail to germinate, or seedlings may be attacked before or after emergence and turn brown, shrink, and finally die. Control measures include these: plant fungicide-treated seeds in well-drained areas; do not apply excessive amounts of nitrate forms of nitrogen fertilizers; and rotate the location of the vegetables. Root Rot of Beans and Southern Peas-Root rot is severe on green beans, lima beans, and southern peas. The disease first appears as reddish or reddish-brown areas on stems and roots. As the disease advances, discolored areas spread until the entire root and lower stem are affected. Above-ground symptoms include stunting, yellowing, drooping of leaves, failure to produce normal pods, and death.

Accurate disease identification is the first step in planning an effective and efficient disease control program. Literally hundreds of different vegetable diseases exist. Although a vegetable grower will encounter only some of these during one season, he must be able to identify them and to distinguish significant and potentially serious diseases from less important diseases. The aim of this book is to help vegetable growers identify important disease.

I would like to thank my team for standing beside me throughout my career and writing this book. My special thanks go to "Random Publications" who have published the book.

– Mahendra Kamalvanshi

Contents

1

The Effect of Raw Materials and Production Conditions on Vegetables Auxiliary Raw Materials

Auxiliary raw materials used in and vegetable processing technologies play a major role in the determination of their physical and chemical characteristics, sensory properties and nutritive value.

WATER

Water is one of the essential factors in the activity of the processing centres; according to the final utilisation, water can be classified in three categories:

a. For technological utilisation (when it comes into direct contact with raw materials and enters in the finished product's composition),
b. For steam generators and
c. For receptacle cooling, washing of equipment and general hygiene.

Water for Technological Uses

Water coming into direct contact with the raw materials used for processing (washing, blanching, etc.) or that used as filling liquid of some canned products, must be of drinking water quality in terms of its physico-chemical and microbiological conditions.More important even than fulfilling drinking water standards, water used for these purposes pests certain specific characteristics related to the technological step or the raw material treated during the processing.

When very hard water is used for blanching vegetables some pecto-calcium and pectomagnesium complexes are formed which starts the hardening of vegetable tissues. This process continues over the pasteurization of the finished product.When is processed in sugar syrup, the use of hard water for the syrup preparation could induce the formation of a pectin-sugar-acid gel facilitated by the medium pH and presence of calcium salts.Soft water has negative consequences associated with mineral and hydrosoluble substances and losses during blanching of vegetables. For some specific products such as peeled tomatoes, green beans and fine texture, the addition of calcium salts (chlorure, sulphate, etc.) is employed to correct for low texture.The water hardness is an essential factor when used as filling liquid for canned products; ideally, the hardness of the water should be adapted to the raw material species used for canning. Thus a hardness of 3° is good for beans, 5° to 9° for green peas, green beans, and for vegetables with a tendency to disintegrate should use even harder water.

In the technological process of cucumber and gherkin preservation by natural acidification (lactic fermentation), water hardness has a paramount role. The literature maintains that as far as texture is concerned, the best results are obtained by using very hard water (about 30°); but since high magnesium and salt content has a negative effect on taste, in practice it is recommended that a water hardness of about 10° be used, which gives satisfactory results from both points of view.Oxygen present in water can act as a corrosion factor in metal receptacles but this negative influence can be eliminated by preliminary boiling.

An important factor is pH. Water for canning must be neutral or slightly basic. Acid water plays a major role in corrosion which is evident both on receptacles and on iron or copper equipment, where changes of product colour will be induced. More dangerous is the attack on lead pipes or to the mix used for can sticking and this can render the product toxic. For these reasons an acid water must be neutralised before use.

Water for steam generators

Two main conditions must be fulfilled:

- Hardness has to be as low as possible, even zero, because precipitation of calcium salts can lead to the formation of encrustations ("crusts") in pipes and on equipment walls. For this reason, water treatment is practiced in the majority of installations;
- From a bacteriological point of view, the iron-bacteria must be eliminated with biological filters or oxidising substances. This is necessary in order to avoid iron hydroxide formed during equipment exploitation to deposit on the inner walls of the pipes.

The elimination of iron-bacteria is also of importance for the water used in processing steps.

Water for Receptacle Cooling and General Hygiene

This should be of drinking water standard.

Where this is not available in sufficient quantity, the use of industrial water is acceptable but only for cleaning of production rooms/workshops.

SWEETENERS

Sugar

Sugar is the conventional name applied to sucrose. Physically there is icing, granulated and lump sugar. In and vegetable processing, sugar is used only in its granulated form; this quality must be in the form of uniform crystals, white, shining and completely soluble in water.

Concentration of various sugar solutions can be rapidly measured by refractometre reading or with areometres graduated in various ways: Brix, Baumé, etc. Sugar solubility in water is dependent upon temperature; for example, in order to obtain a saturated solution, one must dissolve 2040 g in one litre of water at 20° C and 4870 g at 100° C. Taking into account this temperature related solubility, in practice the majority of sugar solutions are prepared by heating the water. Water should be as soft as possible because the calcium salts can precipitate on boiling.

Corn syrup (liquid glucose)

Corn syrup is obtained industrially by acid or enzymatic starch hydrolysis, using as starting raw materials maize (corn) or potatoes. In processing, mainly in the production of marmalades, it is possible to use corn syrup. The average composition of this corn syrup is of about 32-40% dextrose (glucose), about 40% dextrins and 18-20% moisture. Sweetening power is 50% compared with sucrose.

In a 10%-20% proportion with sucrose, addition of corn syrup has certain advantages:

a. It improves the shine and texture of marmalade;
b. It prevents "sugaring" defect and
c. It reduces the too sweet taste of finished products obtained with sugar alone.

SALT

Salt is used in order to give to the finished products a specifically salty taste and as a preserving substance. From a chemical point of view the term salt means sodium chloride but in practice the product is never in a pure state. The presence of a significant quantity of magnesium chloride increases the hygroscopicity, gives a bitter taste and can induce corrosion of receptacles.

Table. Physical Characteristics of Sugar Solutions

Specific weight	deg.Bé	Sugar in solution		Boiling
		temperature °C deg.Bx (K/100 g)	g/l	
1.144	18.5	33	377	101.1
1.149	19.0	34	391	101
1.154	19.6	35	404	101.2
1.159	20.1	36	417	101.3
1.164	20.7	37	430	101.3

1.169	21.2	38	444	101.4
1.174	21.8	39	457	101.4
1.179	22.3	40	470	101.5
1.185	22.9	41	486	101.5
1.190	23.4	42	500	101.6
1.195	23.9	43	513	101.6
1.200	24.5	44	527	101.7
1.206	25.0	45	543	101.7
1.211	25.6	46	557	101.8
1.216	26.1	47	571	101.8
1.222	26.6	48	586	101.9
1.227	27.2	49	600	101.9
1.233	27.7	50	616	102.0
1.238	28.2	51	630	102.1
1.244	28.8	52	646	102.2
1.249	29.3	53	660	102.3
1.255	29.8	54	677	102.4
1.261	30.4	55	693	102.5
1.267	30.9	56	709	102.6
1.272	31.4	57	723	102.7
1.278	31.9	58	740	102.8
1.284	32.5	59	757	102.9
1.290	33.0	60	774	103.0
1.296	33.5	61	790	103.2
1.302	34.0	62	807	103.5
1.308	34.5	63	824	103.7
1.314	35.1	64	840	103.9
1.320	35.6	65	857	104.2
1.326	36.1	66	874	104.6
1.332	36.6	67	891	105.0
1.338	37.1	68	909	105.5
1.345	37.6	69	928	106.0
1.351	38.1	70	945	106.5
1.357	38.6	71	962	106.8
1.364	39.1	72	981	107.1
1.370	39.6	73	999	107.4
1 376	40.1	74	1017	107.8
1.383	40.6	75	1037	108.2
1.389	41.1	76	1058	109.0
1.396	41.6	77	1076	110.0
1.402	42.1	78	1094	111.0
1.409	42.6	79	1113	112.0
1.414	43.1	80	1133	113.0

From a microbiological point of view, salt it is not a sterile product but on the contrary contains various micro-organisms, mainly halophil bacteria.

Salt solubility is only slightly influenced by temperature (0.360 kg/1 at 20° C and 0.390 kg/l at 100° C). Correspondence between specific weight and salt content of salt solutions at 15° C is shown in Table given below.

Table. Physical Characteristics of Salt Solutions

deg.Bé	Specific Weight	NaCl Content	
		g/100 g	g/l
1	1.007	1	10
2	1.014	2	20
3	1.022	3	30
4	1.029	4	41
5	1.037	5	52
6	1.045	6	63
7	1.052	7	74
8	1.060	8	85
9	1.067	9	96
10	1.075	10	107
11	1.083	11	119
12	1.091	12	131
13	1.100	13	143
14	1.108	14.2	158
15	1.116	15.5	173
16	1.125	16.7	188
17	1.134	18.0	204
18	1.142	19.0	217
19	1.152	20.0	230
20	1.162	21.2	246
21	1.171	22.4	262
22	1.180	23.6	278
23	1.190	24.8	295
24	1.200	26.0	312
24.5	1.204	26.4	318

Food Acids

Acetic Acid

Acetic acid is in use as solutions of various concentrations which are known under the generic name of vinegar. Vinegar can be obtained:

a. From wine, alcohol, cider, beer, etc. by fermentation;
b. By dilution of acetic acid obtained by dry wood distillation or by synthesis.

From a quality point of view, wine vinegar is preferred, as it has a more pleasant flavour. In order to improve taste, other vinegar types are usually flavoured with spices.In addition to its spicing and flavouring role, vinegar is used and acts as a preservation agent for some vegetables: cucumbers, acidified vegetables, etc.

- Citric acid
- Tartric acid
- Lactic acid

Pectic Preparations

In processing there many preparations and mixes known as "pectin" are used as liquid or powder extracts.

From a practical point of view, the pectic preparations are classified as:

- Strong pectins obtained from apples or citrus peel; this category gives gels rich in sugar;
- Weak pectins which gives gels with low proportion of sugar or even without sugar but with the addition of calcium salts.

The industry uses mainly strong pectins. These preparations are characterised, from a commercial point of view, by the capacity of gelification, expressed in degrees. The degree of gelification represents the quantity of sugar in grams able to be transformed in a standard gel (65% sugar and pH=3) by 1 g pectin.

Intensive Sweeteners

"Calorie-reduced" and "low-calorie foods are widely used and are cornering an increasing share of the market. Sweeteners are making an important contribution to the manufacture of sweet foods in these categories. They make it possible to manufacture sweet products without "sweet" being necessarily synonymous with "high-calorie".

Diabetics need to restrict their intake of sugar and various carbohydrates similar to sugar or avoid them altogether. Sweeteners enable diabetics to enjoy sweet tastes without changing their lifestyle. Sweeteners do not contribute to the development of tooth decay; they do not degrade in the mouth to form the acids which are responsible for caries. Thus sweeteners offer consumers a number of advantageous and favourable properties above and beyond merely reducing calories.

Sunett is the trade mark of Hoechst AG for its high intensity sweetener acesulfame K. As an ingredient, it can be used for sweetening all foods produced industrially or at home, or to produce tabletop sweeteners. Like all other sweeteners, Sunett tastes more intensely sweet than sugar; it is about 200 times sweeter than sugar compared with a dilute stock solution.

Synergism — Sunett is notable for its pronounced synergism with other sweeteners. The synergistic effect leads as quantitative synergism to an intensification of the overall sweetness and as qualitative synergism to an improved taste. The synergism results in a marked intensification of the sweet taste of the blends, which can amount to 30-50% at usual concentrations.

The favourable properties of Sunett, particularly its synergistic behaviour, can be used advantageously in the Sunett-Multi-Sweetener Concept. For high sweetness levels, blends of Sunett with other sweeteners, e.g. aspartame, are particularly favourable in many applications. The synergistic effect of the blend results in a taste that is particularly pleasant and rounded.Alone or in blends with other sweeteners, Sunett is used mainly where only sweetness is important and no other properties of the foods are to be affected. This applies particularly to beverages (syrups and juices, carbonated beverages, etc.).In a number of food products, sweet carbohydrates do not only provide the sweet taste; they have other functions to fulfil; for example, they act as bulking or texturing agents and as preservatives by reducing the water activity. In these types of products, Sunett and other sweeteners cannot be used on their own. They must be combined with other substances which perform the required functions. These may be bulking agents or sugar substitutes (for example polyols: sorbitol, mannitol, xylitol). Sunett can be combined with both groups of substances.

Uses

In carbonated soft drinks blends of Sunett with other sweet-tasting substances are often recommended. Beverages formulated using the Sunett Multi-Sweetener Concept, i.e. blends of Sunett with other sweeteners, are usually preferred because of their particularly well balanced and rounded sweetness profile. The taste of these blends is often superiors to single sweeteners.Sunett is compatible with sugar and other sweet-tasting carbohydrates, both technically and in terms of taste. It can therefore be used in the production of soft drinks with a reduced sugar content. Drinks based on fructose and Sunett are suitable for diabetics. Soft drinks with a reduced sugar content show an excellent taste quality when Sunett or Sunett blends are added to bring them up to the usual level of sweetness. nectars and juice drinks differ from most carbonated drinks in that they contain a noticeable amount of sweet carbohydrates provided by the juice. The amount of carbohydrate may vary depending on the type of and amount of juice used.Amounts of up to 200 mg/l Sunett are often adequate as a single sweetener for the popular types of nectars. With blends of Sunett and aspartame quantities in the order of 100150 mg/l Sunett and 50 mg/l aspartame or 60-70 mg/l of both Sunett and aspartame are sufficient.

Sunett offers such excellent stability that end products, such as drinks, show no reduction in sweetness performance during normal processing methods and storage periods. Sunett can withstand pasteurization, hot filling and aseptic filling without any loss of sweetness.

Jams and marmalades. Sugar contributes a great deal to the texture and stability of conventional jams and marmalades. For the production of sugar-free products either the sugar must be replaced by comparable amounts of sugar substitutes or some of the functions of the sugar must be taken only by other components, such as suitable gelling agents.

Sugar-free jams and marmalades containing sweeteners are more susceptible to microorganisms than sugar-containing products. The risk of spoilage due to yeast fermentation or moulding can be prevented by pasteurization. However, this is only feasible for small jars which will be quickly consumed once they are opened.In all other cases, it is advisable to add 0.05-0.1% potassium sorbate as a preservative, wherever this is permitted under the relevant food regulations. For sugar-free jams and marmalades, concentrations in the range of 500-2000 mg Sunett/kg of the finished product are appropriate. It is advisable to add Sunett in the form of an aqueous stock solution towards the end of the boiling process. Care must be taken to ensure that Sunett is evenly dispersed throughout the whole batch.Because of the excellent compatibility of Sunett with sugar alcohols, jams and marmalades using these ingredients offer an outstanding taste.

Preserves

Sunett can be used for the production of sugar-free or sugar-reduced preserves. At the normal pH values for preserves Sunett can be added even before pasteurization, as the sweetness is not impaired under the usual thermal treatment conditions. Sunett also withstands the usual storage periods without loss of sweetness.

NutraSweet(r) is the commercial name of aspartame (APM), a new sweetener from G.D. SEARLE & Co. which can be used in the most foods in order to give the same taste as sugar. NutraSweet is about 180 to 200 times sweeter than sucrose (sugar) and this value depends on pH, temperature and the type of flavour.NutraSweet can be used as mentioned above in a mix with Sunett or alone in all sugar-free or calorie-reduced jams, marmalades and preserves. Like other sweeteners, NutraSweet does not promote tooth decay.

International Regulatory Status

Both Sunett and NutraSweet are widely accepted by food laws in the majority of countries. The following are the main products where Sunett is an accepted sweetener:

- Low-joule prepared jelly: max. 500 mg/kg;

- Canned without added sugar: max. 500 mg/kg;
- Beverages including calorie-reduced nectars: max. 600 mg/l;
- Calorie-reduced jams;
- Calorie-reduced and dietetic compotes;
- Canned and vegetables, puree, jams and marmalades;
- Juices, nectars and juice based beverages;
- Jams, marmalades and related products: max. 300 mg/kg.

77

Packaging MaterialsPackaging MaterialsRequirements and functions of food containers

The following are among the more important general requirements and functions of food packaging materials/containers:

a. They must be non-toxic and compatible with the specific foods;
b. Sanitary protection;
c. Moisture and fat protection;
d. Gas and odour protection;
e. Light protection;
f. Resistance to impact;
g. Transparency;
h. Tamperproofness;
i. Ease of opening;
j. Pouring features;
k. Reseal features;
l. Ease of disposal;
m. Size, shape, weight limitations;
n. Appearance, printability;
o. Low cost;
p. Special features.

PRIMARY AND SECONDARY CONTAINERS

The terms primary and secondary containers have been used. Some foods are provided with efficient primary containers by nature, such as nuts, oranges, eggs and the like. In packaging these, we generally need only a secondary outer box, wrap, or drum to hold units together and give gross protection.

Other foods such as milk, dried eggs and concentrates often will be filled into primary containers such as plastic liners which are then packaged within protective cartons or drums. In this case the secondary container provided by the carton or drum greatly minimises the requirements that must be met by the primary container.

Except in special instances, secondary containers are not designed to be highly impervious to water vapour and other gases, especially at zones of sealing, dependence for this being placed upon the primary container. Since primary containers by definition are those which come in direct contact with the food, we will be far more concerned with them than with secondary containers.

Hermetic Closure

Two conditions of the greatest significance in packaging are hermetic and non-hermetic closure. The term hermetic means a container which is absolutely impermeable to gases and vapours throughout its entirety, including its seams.

Such a container, as long as it remains intact, will automatically be impervious to bacteria, yeasts, moulds, and dirt from dust and other sources since all of these agents are considerably larger than gas or water vapour molecules.On the other hand, a container which prevents entry of micro-organisms, in many instances will be non-hermetic. A container that is hermetic not only will protect the product from moisture gain or loss, and from oxygen pickup from the atmosphere, but is essential for strict vacuum and pressure packaging.The most common hermetic containers are rigid metal cans and glass bottles, although faulty closures can make them non-hermetic. With very rare exceptions flexible packages are not truly hermetic for one or more of the following reasons.

First, the thin flexible films, even when they do not contain minute pinholes, generally are not completely gas and water-vapour impermeable although the rates of gas and water vapour transfer may be exceptionally slow; second, the seals are generally good but imperfect; and third, even where film materials may be gas- and water-vapour-tight, such as certain gages of aluminium foil, flexing of packages and pouches leads to minute pinholes and crease holes.Hermetic rigid aluminium containers can be readily formed without side seams or bottom end seams. The only seam then to make hermetic is the top end double seam, which may be closed on regular tin can sealing equipment.Glass containers are hermetic provided the lids are tight. Lids will have inside rings of plastic or cork. Many glass containers are vacuum packed and the tightness of the cover will be augmented by the differential of atmospheric pressure pushing down the cover. Crimping of the covers, as in the case of pop bottle caps which operate against positive internal pressure, also can make a gas-tight hermetic seal. But bottles fail more often than cans in becoming non-hermetic.

Protection of Food by Packaging Materials

Important factors in selecting a packaging unit for food storage are presented in Fig.

Fig. Factors for Selection a Packaging Material for Food Storage

Climatic Hazards	Biological Hazards
Atmospheric Humidity	Insects
Light (UV)	Bacteria & Moulds

Temperature	Mites	
Rain	Rodents & Birds	
	Product	
Handling	Chemical Composition	Types of Stores Customer Appeal Transport
Stacking	Price	

Films and Foils; Plastics

Films and foils have different values for moisture and gas permeability, strength, elasticity, inflammability and resistance to insect penetration and many of these characteristics depend upon the film's thickness.

Important characteristics of the types of films and foils commonly used in food packaging are given in Table given below. For the most part such films are used in the construction of inner containers. Since they are non-rigid, their main functions are to contain the product and protect it from contact with air or water vapour. Their capacity to protect against mechanical damage is limited, particularly when thin films are considered.

Table. Properties of Packaging Films

Material	Properties
Paper	Strength; rigidity; opacity; printability.
Aluminium foil	Negligible permeability to water-vapour, gases and odours; grease proof, opacity and brilliant appearance; dimensional stability; dead folding characteristics.
Cellulose film (coated)	Strength; attractive appearance; low permeability to water vapour (depending on the type of coating used), gases, odours and greases; printability.
Polythene	Durability; heat-sealability; low permeability to water-vapour; good chemical resistance; good low-temperature performance.
Rubber hydrochloride	Heat-sealability; low permeability to water vapour, gases, odours and greases; chemical resistance.
Cellulose acetate	Strength; rigidity; glossy appearance; printability; dimensional stability.
Vinylidene chloride	Low permeability to water vapour, gases, copolymer odours and greases; chemical resistance; heat-sealability.
Polyvinyl chloride	Resistance to chemicals, oils and greases; heat-sealability.
Polyethylene terephtalate	Strength; durability; dimensional stability; low permeability to gases, odours and greases.

These materials can exist in many forms, depending upon such variables as identity and mixture of polymers, degree of polymerisation and molecular weight, spatial polymer orientation, use of plasticisers (softeners) and other chemicals, methods of forming such as casting, extrusion or calendering, etc.

One of the newer classes of plastic materials referred to as copolymers illustrate what can be done with mixtures of the basic units from which plastics are built. The term copolymer refers to a mixture of chemical species in the resin from which films and other forms can be made. The many variations possible make copolymers an important class of plastics to extend the range of useful food packaging applications.

Plastic Sheets

- Cellophane paper can be used for packing of dried products, mainly for dried leathers.
- Polyethylene sheets have a variety of uses. They are flexible, transparent and have a perfect resistance to low temperatures and impermeability to water vapour. An important advantage is that these sheets can be easily heat-sealed. Utilisation is in forms of sheets and bags. It is a good packing material for primary protection of dehydrated products. If a good protection is needed to prevent flavour and gas losses, it will be necessary to combine polyethylene with other materials.

Receptacles and Packagings in Plastic Materials

In this class there are three categories:

a. Receptacles that can be heat treated: boxes, bottles and bags. Sterilisable bags used up to 120° C can be manufactured from same raw materials as described under plastic sheets and up to 100° C from cellophane. Polyethylene bags could be used to some extent for packing and pasteurization of sauerkraut.
b. Receptacles that are not heat treated during processing of and vegetables, also divided in bags and boxes. Bags are the most used type of packing from plastic materials and they are manufactured from polyethylene or cellophane; an important utilisation is for dried/ dehydrated s and vegetables.
c. special packagings - which can be contacted (Criovac type) by action of heat once the finished product is already inside the pack and the air is evacuated.

Laminates

Various flexible materials such as papers, plastic films, and thin metal foils have different properties with respect to water vapour transmission, oxygen permeability, light transmission, burst strength, pin holes and crease hole sensitivity, etc. and so multi-layers or laminates of these materials which combine the best features of each are used. Commercial laminates containing up to as many as eight different layers are commonly custom designed for a particular product.

Laminations of different materials may be formed by various processes including bonding with a wet adhesive, dry bonding of layers with a thermoplastic adhesive, hot melt laminating where one or both layers exhibit thermoplastic properties, and special extrusion techniques. Such structured plastic films may be complete in themselves or be further bonded to papers or metal foils to produce more complex laminates.

GLASS CONTAINERS

As far as food packaging is concerned, glass is chemically inert, although the usual problems of corrosion and reactivity of metal closures will of course apply. The principal limitation of glass is its susceptibility to breakage, which may be from internal pressure, impact, or thermal shock, all of which can be greatly minimised by proper matching of the container to its intended use and intelligent handling practices. Here consultation with the manufacturer cannot be over-stressed.

The heavier a jar or bottle for a given volume capacity the less likely it is to break from internal pressure. The heavier jar, however, is more susceptible to both thermal shock and impact breakage. Greater thermal shock breakage of the heavier jar is due to wider temperature differences which cause uneven stress between the outer and inner surfaces of the thicker glass. Greater impact breakage susceptibility of the heavier jar is due to the lower resilience of its thicker wall.

Coatings of various types can markedly reduce each of these types of breakage. These coatings, commonly of special waxes and silicones, lubricate the outside of the glass. As a result, impact breakage is lessened by bottles and jars glancing off one another rather than sustaining direct hits when they are in contact in high speed filling lines.Surface coating after annealing protects glass surfaces from many of the minute scratches appearing in normal handling after annealing ovens; surface coating also improves the high gloss appearance of glass containers and is said to decrease the noise from glass to glass contact at filling lines.With regard to thermal shock, it is good practice to minimise temperature differences between the inside and outside of glass containers whenever possible. Some manufacturers will recommend that a temperature difference of 44° C (80° F) between the inside and the outside not be exceeded. This requires slow warming of bottles before use for a hot fill and partial cooling before such containers are placed under refrigeration.

Classification

Glass used for receptacles in and vegetable processing is a carefully controlled mixture of sand, soda ash, limestone and other materials made molten by heating to about 1500° C (2800° F).

Main classes of glass receptacles are:

a. Jars which are resistant to heat treatments,
b. Jars, glasses, etc. for products not submitted to heat treatment (marmalades, acidified vegetables, etc.);
c. Glass bottles for pasteurized products (tomato juice, juices, etc.) or not pasteurized (syrups) and
d. Receptacles with higher capacity (flasks, etc.)

Jars for Sterilised/Pasteurized Canned Products

These receptacles may replace metal cans, taking into considering both the advantages and disadvantages they present. Advantages are: they do not react to food content; they are transparent and can be manufactured in various shapes; they use cheap raw materials and are reusable. Disadvantages: heavier than metal can of same capacity; fragile; lower thermal conductance and a limited resistance to thermal shocks.

Receptacles in this category must assure a perfect hermeticity after their pasteurization/sterilisation and cooling and this has to be achieved by the use of metallic (or glass) caps and specific materials for tightness. Taking into account the receptacles' closure method, there are two categories of receptacles:

a. Glass jars with mechanical closure;
b. Glass jars with pneumatic closure;

Jars for Products without Heat Treatment

For marmalades, jellies and jams glass jars with non hermetic closures using metal, glass or rigid plastic caps are used; however for these products the receptacles mentioned above may also be used.The use of jars with pneumatic closure presents the advantage that some products (e.g. marmalades, jams) can be filled hot and therefore sterile in receptacles. Pneumatic closing generally protects against negative air action which is in this case eliminated from receptacles.

Glass Bottles

These receptacles are widely used both for

a. Finished products which need pasteurization (ea. tomato juice, Knit juices, etc.) and for
b. Those which are preserved as such (ea. syrups).

Glass bottles in category a) are closed hermetically with metallic caps, provided with special materials for tightness. For glass bottles in category b) various corks, and aluminium caps with tightness materials may be used.

Glass Receptacles with High Capacity

In this category there are glass flasks with 3 and 10 litre capacity which can be hermetically closed by a SKO caps system and are resistant to product pasteurization (ea. tomato juice). As bigger receptacles it is possible to use glass demijohns with usual capacity of 25 and 50 litre; these receptacles are used for preservation of juices by warm process. Closing is performed with flexible rubber hoods.

Paper Packaging

As primary containers few paper products are not treated, coated or laminated to improve their protective properties. Paper from wood pulp and reprocessed waste paper will be bleached and coated or impregnated with such materials as waxes, resins, lacquers, plastics, and laminations of aluminium to improve water vapour and gas impermeability, flexibility, tear resistance, burst strength, wet strength, grease resistance, sealability, appearance, printability, etc.

Paper Sheets

- Kraft paper is the brown unbleached heavy duty paper commonly used for bags and for wrapping; it is seldom used as a primary container;
- Parchment paper: acid treatment of paper pulp modifies the cellulose and gives water and oil resistance and considerable wet strength to this type of packaging material;
- Glassine-type papers are characterised by long wood pulp fibres which impart increased physical strength;
- Paper with plastic material sheets.

Receptacles from Paper or Cardboard

(paper = 8 to 150 g/m^2; cardboard = 150 to 450 g/m^2).

"Tin can"/Tinplate

The "tin can" is a container made of tinplate. Tinplate, a rigid and impervious material, consists of a thin sheet of low carbon steel coated on both sides with a very thin layer of tin. It can be produced by dipping sheets of mild steel in molten tin (hot-dipped tinplate) or by the electro-deposition of tin on the steel sheet (electrolytic tinplate). With the latter process it is possible to produce tinplate with a heavier coating of tin on one surface than the other (differentially coated).

Tin is not completely resistant to corrosion but its rate of reaction with many food materials is considerably slower than that of steel. The effectiveness of a tin coating depends on:

a. Its thickness which may vary from about 0.5 to 2.0 μm (20 to 80 x 10(-6) in.);
b. The uniformity of this thickness;
c. The method of applying the tin which today primarily involves electrolytic plating;
d. The composition of the underlying steel base plate;
e. The type of food, and
f. Other factors.

Some canned vegetables including tomato products actually owe their characteristics flavours to a small amount of dissolved tin, without which these products would have an unfamiliar taste. On the other hand, where tin reacts unfavourably with a particular food the tin itself may be lacquer coated.The thickness of tinplate sheets may vary from 0.14 mm to 0.49 mm and is determined by weighing a sheet of known area and calculating the average thickness.

Tinplate sheets may be lacquered after fabrication to provide an internal or external coating to protect the metal surface from corrosion by the atmosphere or through reaction with the can contents. They may also be printed by lithography to provide suitable instructions or information on containers fabricated from tinplate sheets (otherwise paper labels can be attached to the outer tinplate surface).

Under normal conditions the presence of the tin coating provides a considerable degree of electrochemical protection against corrosion, despite the fact that in both types of tinplate the tin coating is discontinuous and minute areas of steel base plate are exposed. With prolonged exposure to humid conditions, however, corrosion may become a serious problem.

The coatings not only protect the metal from corrosion by food constituents but also protect the foods from metal contamination, which can produce a host colour and flavour reactions depending upon the specific food. Particularly common are dark coloured sulphides of iron and tin produced in low acid foods that liberate sulphur compounds when heat processed, and bleaching of red plant pigments in contact with unprotected steel, tin, and aluminium.

Table. General Types of can Coatings

Coating	Typical uses	Type	
enamel	Dark coloured berries, cherries and other s requiring protection from metallic salts	Oleoresinous	
C-enamel	Corn, peas and other sulphur-bearing products	Oleoresinous w. suspended zinc	oxide
Citrus enamel	Citrus products and concentrates oleoresinous	Modified	
Beverage can enamel	Vegetable juices; red juices; highly corrosive s; non-carbonated	Two-coated w. resinous base coat beverages and	vinyl top coat

SPECIFIC PRESERVATION TECHNOLOGIESFRUIT SPECIFIC PRESERVATION TECHNOLOGIES

QUALITY

quality goes back to tree stock, growing practices and weather conditions. Closer to the shipper and processor, however, are the degrees of maturity and ripeness when picked and the method of picking or harvesting.There is a distinction between maturity and ripeness of a. Maturity is the condition when the is ready to eat or if picked will become ready to eat after further ripening. Ripeness is that optimum condition when colour, flavour and texture have developed to their peak.

Some is picked when it are mature but not yet ripe. This is especially true of very soft like cherries and peaches, which when fully ripe are so soft as to be damaged by the act of picking itself. Further, since many types of continue to ripen off the tree, unless they were to be processed quickly, some would become overripe before they could be utilised if picked at peak ripeness.

From a technological point of view, characterisation by species and varieties is performed on the basis of physical as well chemical properties: shape, size, texture, flavour, colour/pigmentation, dry matter content (soluble solids content), pectic substances, acidity, vitamins, etc. These properties are directly correlated with utilisation. Table given below shows which of the above mentioned properties have a major impact on the finished products obtained by processing.

Table. Optimal Use of s as a Function of their Properties

Processed Finished Products	Organoleptic (Sensory) Properties				Chemical Composition		
	Shape	Texture	Flavour	Taste Sust.	Acidity	Sugars	Pectic
Dried s	+ +	+ +		+ +		+ +	
Juices			+ +	+ +	+ +		
Marmalade			+ +	+ +			+ +
Jams	+ +	+ +	+ +	+ +			
Jellies	+ +	+ +	+ +	+ +			+ +
Paste				+ +		+ +	

The proper time to pick depends upon several factors; these include variety, location, weather, ease of removal from the tree (which change with time), and purpose to which the will be put.

For example, oranges change with respect to both sugar and acid as they ripen on the tree; sugar increases and acid decreases. The ratio of sugar to acid determines the taste and acceptability of the and the juice. For this reasons, in some countries there are laws that prohibit picking until a certain sugar-acid ratio has been reached.

In the case of much to be canned, on the other hand, is picked before it is fully ripe for eating since canning will further soften the.

QUALITY MEASUREMENTS

Many quality measurements can be made before a crop is picked in order to determine if proper maturity or degree of ripeness has developed. Colour may be measured with instruments or by comparing the colour of on the tree with standard picture charts. Texture may be measured by compression by hand or by simple type of plungers.

As mature on the tree its concentration of juice solids, which are mostly sugars, changes. The concentration of soluble solids in the juice can be estimated with a refractometre or a hydrometre. The refractometre measures the ability of a solution to bend or refract a light beam which is proportional to the solution's concentration. A hydrometre is a weighted spindle with a graduated neck which floats in the juice at a height related to the juice density.

The acid content of changes with maturity and affects flavour. Acid concentration can be measured by a simple chemical titration on the juice. But for many s the tartness and flavour are really affected by the ratio of sugar to acid.Percentage of soluble solids, which are largely sugars, is generally expressed in degrees Brix, which relates specific gravity of a solution to an equivalent concentration of pure sucrose.

In describing the taste of tartness of several s and juices, the term "sugar to acid ratio" or "Brix to acid ratio" are commonly used. The higher the Brix the greater the sugar concentration in the juice; the higher the "Brix to acid ratio" the sweeter and lees tart is the juice.

HARVESTING AND PREPROCESSING

Harvesting

The above and other measurements, plus experience, indicate when is ready for harvesting and subsequent processing. A large amount of the harvesting of most crops is still done by hand; this labour may represent about half of the cost of growing the. Therefore, mechanical harvesting is currently one of the most active fields of research for the agricultural engineer, but also requires geneticists to breed of nearly equal size, that matures uniformly and that is resistant to mechanical damage.

A correct manual harvesting includes some simple but essential rules:

- The should be picked by hand and placed carefully in the harvesting basket; all future handling has to be performed carefully in order to avoid any mechanical damage;
- The harvesting basket and the hands of the harvester should be clean;
- The should be picked when it is ready to be able to be processed into a quality product depending on the treatment which it will undergo.

It is worth emphasising the fact that the proximity of the processing centre to the source of supply for fresh raw materials presents major advantages; some are as follows:

- Possibility to pick at the best suitable moment;
- Reduction of losses by handling/transportation;
- Minimises raw material transport costs;
- Possibility to use simpler/cheaper receptacles for raw material transport.

Once it has left the tree, the organoleptic properties, nutritional value, safety and aesthetic appeal of the deteriorates in varying degrees. The major causes of deterioration include the following:

a. Growth and activity of micro-organisms;
b. Activities of the natural food enzymes;
c. Insects, parasites and rodents;
d. Temperature, both heat and cold;
e. Moisture and dryness;
f. Air and in particular oxygen;
g. Light and
h. Time.

The rapidity with which foods spoil if proper measures are not taken is indicated in Table given below.

Table. Useful Storage Life of Some Food Products

Food Products	Generalized Storage Life (Days) at 21°C (70°F)
Animal Flesh, Fish, Poultry	1-2
Dried, salted, smoked meat and fish	360 and more
s	1-7
Dried s	360 and more
Leafy Vegetables	1-2
Root Crops	7-20

Reception — Quality and Quantity

Reception at the processing centre is performed mainly for following purposes:

- Checking of sanitary and freshness status;
- Control of varieties and wholeness;
- Evaluation maturity degree;
- Collection of data about quantities received in connection to the source of supply: outside growers/farmers, own farm.

Variety control is needed in order to identify that the belongs to an accepted variety as not all are suitable for different technological processes. maturity degree is significant as industrial maturity is required for some processing/ preservation methods while for others there is the need for an edible maturity when the has full taste and flavour.

Special attention is given to size, appearance and uniformity of to be processed, mainly in the form of preserved with sugar using whole/half s ("with pieces"). Some laboratory control is also needed, even if it not easy to precisely establish the technological qualities of because of the absence of enough reliable rapid analytical methods able to show eventual deterioration.

The only reliable method for evaluating the quality is the combination of data obtained through organoleptic/taste controls and by simple analytical checks which are possible to perform in a small laboratory: percentage of soluble solids by refractometre, consistency/texture measured with simple penetrometres, etc.

Temporary Storage before Processing

This step has to be as short as possible in order to avoid flavour losses, texture modification, weight losses and other deterioration that can take place over this period.

Some basic rules for this step are as follows:

- Keep products in the shade, without any possible direct contact with sunlight;
- Avoid dust as much as possible;
- Avoid excessive heat;
- Avoid any possible contamination;
- Store in a place protected from possible attack by rodents, insects, etc.

Cold storage is always highly preferred to ambient temperature. For this reason a very good manufacturing practice is to use a cool room for each processing centre; this is very useful for small and medium processing units as well.

Table. Raw Material Control — Fresh s and Vegetables at Reception

1.	Checks at each delivery/raw material lot Colour, Texture, Taste, Flavour, Appearance, Refractometric, extract, Number per kg, Variety, Sanitary evaluation
2.	Checks at each ten lots (for the same raw material) Density, Water content: oven method, Total sugars, reducing sugars, Total acidity
3.	Audits — every six months — on five different lots Ascorbic acid, Mineral substances, Tannic substances, Pectic substances.

The type of analysis for audits will be adapted to the specific s and vegetables that are received/processed. An excellent indication of a good temporary storage is the limited weight loss before processing, which has to be below 1.0%-1.2%.

Washing

Harvested is washed to remove soil, micro-organisms and pesticide residues. washing is a mandatory processing step; it would be wise to eliminate spoiled before washing in order to avoid the pollution of washing tools and/or equipment and the contamination of during washing. Washing efficiency can me gauged by the total number of micro-organisms present on surface before and after washing - best result are when there is a six fold reduction. The water from the final wash should be free from moulds and yeast; a small quantity of bacteria is acceptable. washing can be carried out by immersion, by spray/ showers or by combination of these two processes which is generally the best solution: pre-washing and washing.

Some usual practices in washing are:

- Addition of detergents or 1.5% HCl solution in washing water to remove traces of insect-fungicides;
- Use of warm water (about 50°C) in the pre-washing phase;
- Higher water pressure in spray/shower washers.

Washing must be done before the is cut in order to avoid losing high nutritive value soluble substances (vitamins, minerals, sugars, etc.).

Sorting

Sorting covers two main separate processing operations:

a. Removal of damaged and any foreign bodies (which might have been left behind after washing);
b. Qualitative sorting based on organoleptic criteria and maturity stage.

Mechanical sorting for size is usually not done at the preliminary stage. The most important initial sorting is for variety and maturity.

However, for some and in special processing technologies it is advisable to proceed to a manual dimensional sorting (grading).

Trimming and Peeling (Skin Removal)

This processing step aims at removing the parts of the which are either not edible or difficult to digest especially the skin.

Up to now the industrial peeling of and vegetables was performed by three procedures:

a. Mechanically;
b. By using water steam;
c. Chemically; this method consists in treating and vegetables by dipping them in a caustic soda solution at a temperature of 90 to 100° C; the concentration of this solution as well as the dipping or immersion time varying according to each specific case.

Cutting

This step is performed according to the specific requirements of the processing technology.

Heat Blanching

Is not usually heat blanched because of the damage from the heat and the associated sogginess and juice loss after thawing. Instead, chemicals are commonly used without heat to inactivate the oxidative enzymes or to act as antioxidants and they are combined with other treatments.

Ascorbic/Citric Acid Dip

Ascorbic acid or vitamin C minimises oxidation primarily by acting as an antioxidant and itself becoming oxidised in preference to catechol-tannin compounds. Ascorbic acid is frequently used by being dissolved in water, sugar syrup or in citric acid solutions.It has been found that increased acidity also helps retard oxidative colour changes and so ascorbic acid plus citric acid may be used together. Citric acid further reacts with (chelates) metal ions thus removing these catalysts of oxidation from the system.

Sulphur Dioxide Treatment

Sulphur dioxide may function in several ways:

- Sulphur dioxide is an enzyme poison against common oxidising enzymes;
- It also has antioxidant properties; i.e., it is an oxygen acceptor (as is ascorbic acid);
- Further SO2 minimises non enzymatic Maillard type browning by reacting with aldehyde groups of sugars so that they are no longer free to combine with amino acids;
- Sulphur dioxide also interferes with microbial growth.

In many processing pre-treatments two factors must be considered:

a. Sulphur dioxide must be given time to penetrate the tissues;
b. SO2 must not be used in excess because it has a characteristic unpleasant taste and odour, and international food laws limit the SO2 content of products, especially of those which are consumer oriented (e.g. except semi-processed products oriented to further industrial utilisation).

Commonly a 0.25 % solution (except for semi-processed products which are industry oriented and use a 6% solution) of SO2 or its SO2 equivalent in the form of solutions of sodium sulphite, sodium bisulphite or sodium/potassium metabisulphite are used. slices are dipped in the solution for about two to three minutes and then removed so as not to absorb too much SO2. Then the slices are allowed to stand for about one to two hours so that the SO2 may penetrate throughout the tissues before processing. Sulphur dioxide is also used in juice production to minimise oxidative changes where relatively low heat treatment is employed so as not to damage delicate juice flavour.Dry sulphuring is the technological step where is exposed to fumes of SO2 from burning sulphur or from compressed gas cylinders; this treatment could be used in the preparation of s (and some vegetables) prior to drying/dehydration.

Sugar Syrup

Sugar syrup addition is one of the oldest methods of minimising oxidation. It was used long before the causative reactions were understood and remains today a common practice for this purpose.

Sugar syrup minimises oxidation by coating the and thereby preventing contact with atmospheric oxygen. Sugar syrup also offers some protection against loss of volatile esters and it contributes sweet taste to otherwise tart s. It is common today to dissolve ascorbic acid and citric acid in the sugar syrup for added effect or to include sugar syrup after an SO2 treatment.

Fresh Storage

Some species and specially apples and pears can be stored in fresh state during cold season in some countries' climatic conditions. for fresh storage have to be autumn or winter varieties and be harvested before they are fully mature. This also has to be sound and without any bruising; control and sorting by quality are mandatory operations.

Sorting has to be carried out according to size and weight and also by appearance; which is not up to standard for storage will be used for semi-processed product manufacturing which will be submitted further to industrial processing. Harvested has to be transported as soon as possible to storage areas. Leaving in bulk in order to generate transpiration is a bad practice as this reduces storage time and accelerates maturation processes during storage. In order to store large quantities of, silos have to be built drying and dehydration technology

General technical data for dehydration in tunnels are presented in Table given below.

Table. Technical Data for Dehydration in Tunnels

s	Drying Conditions			Finished Product	
	Load kg/m²	Temperature °C	Time	Moisture %	Yield %
Plums	15	I. 40-50 II. 75-80	6 H 14 H	18-20	25-35
Apples (Rings)	10	75-55	5-6 H	20	10-12
Apricots (Halves)	10	70-60	10- 15	15-20	10-15
Cherries (w. stones)	10	55-70	6-8	12-15	25
Pears (Halves and quarters)	15	70-65	15-22	18-20	10-15
	15	70-60	10-15	15-20	10-15

For with a high sugar content drying temperatures have to be lower at initial stage and then increase to the maximum acceptable; for with lower sugar level the temperatures are applied in a reverse order.

Some pre-treatments of and vegetables for sun/solar drying are described in table given above.

Table. Technical Data on some Osmotically Dehydrated Products

or vegetable	Type of cut	Treatment	
Banana	5 mm slices 70°C	2 hours, 80% sugar 2000 ppm	SO_2 at
Carrots	10 x 10 x 2 mm dices or 5 mm slices	4 hours, 60% sugar + 10% salt 4000 ppm SO_2	
Mango, green	8 mm slices	2 hours, 25% salt 8000 ppm SO_2	
Mango, ripe	8 mm slices	2 hours, 60% sugar 8000 ppm	SO_2
Onions	2 mm slices	2 hours, 60% sugar + 10% salt 4000 ppm SO2	
Papaya 70°C	8 x 8 mm slices	4 hours, 80% sugar 2000 ppm	SO_2 at
Strawberries	Whole	4 hours, 80% sugar 4000 ppm SO_2	
Sweet peppers, red	6 mm dices	2 hours, 60 % sugar + 10 % salt 4000 ppm SO_2	

Table. Moisture and Shipping Factors for some Dried/Dehydrated s

Products	Form	Moisture, %
Apples	6 nun rings	20
Apricots	Caps	17-20
Banana	Cut pieces	15
Cherries	Whole	12-15
Figs	Whole	23
Guava	Quarters	6
Mango	15 mm pulp sheets	15
Peaches	Caps	15-20
Pears	Halves	23
Prunes	Whole	18-20
Raisins	Whole	17

The moisture contents listed are considered as the best from a technical, practical and commercial point of view for delivery to the market or for shipping and safe for the shelf life needed before buying/consumption by customers/ consumers. All instructions about packing, storage and transport must be followed in order to assure delivery of a safe and high quality product to the market.

Processing of Bars

The bar processing method developed for FAO only involves a single major operation, which it drying the pulp after mixing it with suitable ingredients. It can be used to produce mango, banana, guava or mixed bars.

A dual-powered dryer, working by solar energy during the day and by electric or steam power at night and on rainy days, with cross-flow movement of air and controlled temperature (from 55° C at the beginning of processing to a high of 70° C), is well suited for dehydration of the pulp to the desired moisture level of 15 to 20%.

Main raw material quantities to prepare approximately 100 kg of bars are as follows:

Type of	required in kg	Pulp obtained in kg	Sugar required in kg	Yield (% of fresh) approx.
Mango	720	360	33	14
Banana	600	360	30	17
Guava	406	325	60	25
Mango + banana	540 + 150	360	35	15
Papaya + banana	500 + 140	336	54	23

Mango bar — Fully ripe mangoes are selected and washed in water at room temperature. The peeled is cut into slices and passed through a helicoidal pulper to extract the pulp. The required amount of sugar to adjust the Brix (the unit measure for total solids in s) of the mixed pulp to 25 degrees Brix is then added.Two grams of citric acid per kilogram of pulp (or 20 ml of lime or lemon juice) are added to inhibit possible growth of micro-organisms during drying. The mixture is then heated for two minutes at 80° C and partially cooled; the heat treatment serves to inactivate the enzymes and destroy the micro-organisms.Potassium or sodium metabisulphite is added (two grams per kg of prepared mixture), so that the concentration of SO2 is 1000 ppm. The mixture is then transferred to stainless steel trays which have been previously smeared with glycerine (40 ml/m^2). Each tray must be loaded with 12.5 kg of mixture per square metre.

Drying could be carried out by a dual-powered dryer for a total of 26 hours:

a. 10 hours by solar energy at about 55° C and
b. 16 hours by electric or steam power at 70° C.

At the end of the drying operation, when moisture content is between 15 and 20%, the pieces of suitable shape and size are wrapped in cellophane paper, packed in cartons and stored at ambient air temperature. Pieces of unsuitable shape and size are further cut into small pieces and used to prepare, along with peanuts and cashews, a variety of "cocktail mixtures".

Banana bar. - Banana varieties which give smooth pulp without serum separation must be used for this purpose. Ripe, suitable is selected. The hand-peeled s are soaked in 0.3 per cent citric acid solution for about 10 minutes (lime or lemon juice can replace citric acid). The drained are pulped to obtain smooth pulp.The rest of the procedure is the same as in the case of the mango bar.

Guava Bar

A mixture of pink and yellow varieties is best suited for preparing the bar. The washed is hand peeled and stem and blossom ends trimmed. The peeled is cut into quarters which are passed through a helicoidal extractor to separate seeds and fibrous pieces (the holes in the stainless steel screen should be between 0.8 and 1.10 mm).To get the maximum yield of pulp, the material is passed through the extractor twice. After adjusting the refractometric solids to 25 degrees Brix, the bar can be prepared by following the same procedure as for mango pulp.

Mixed Bar

Mango and banana pulp, as well as papaya and banana pulp, can be mixed in a calculated ratio for preparing mixed bars. The rest of the procedure is the same as in the case of pure mango pulp.

Packing and Storage

The dried pulp is removed from the dryer and cut into square pieces of 5 x 5 cm at a thickness of about 0.3 cm. These pieces, arranged in three layers make up blocks of about 0.9 cm thickness weighing between 25 and 28 grams. An unit pack consist of two such blocks and weights between 50 and 56 grams.

Each block is separately wrapped in cellophane and the unit pack is filled in a printed cellophane bag of size 15 x 8 cm. Two hundred unit packs are packed in a master carton of size 34 x 22 x 14 cm, with a net weight of about 10 kg. Shelf-life is about one year at room temperature.

Leathers

leathers are manufactured by drying/dehydration of purées into leathery sheets. The leathers are eaten as confections or cooked as a sauce. They are made from a wide variety of s, the more common being apple, apricot, banana, cherry, blackcurrant, grape, peach, pear, pineapple, plum, raspberry, strawberry, kiwi, mango and papaya.A description of procedures for mango, banana, guava and mixed bars is given in this document. Another product with good potential is ciku leather; ciku is grown in Malaysia.

A standard process is carried out using ripe s which are washed, peeled, diced and the seeds removed. The s are blanched for 1 minute at 80° C and blended into puree in a food processor.Ciku leather is prepared by mixing ciku puree with 10% sugar, 10% pre-gelatinous rice flour, 150 ppm sorbic acid an 500 ppm sodium metabisulphite (Na2H2SO4). The mixture is cooked on a water bath at 60° C and then made into sheets 1.8 mm thick on trays spread with glycerol to reduce stickiness. This is then further dried in a forced-air dehydrator at 45° C for 3.5 hr or until the surface no longer feels sticky when touched with the fingers. The dried and cooled leathers are cut into 12 x 12 cm squares and wrapped in polypropylene (PP) of 0.1 mm thickness.

Osmotic Dehydration in and Vegetable Processing

Osmotic dehydration is a useful technique for the concentration of and vegetables, realised by placing the solid food, whole or in pieces, in sugars or salts aqueous solutions of high osmotic pressure. It gives rise to at least two major simultaneous counter-current flows: a significant water flow out of the food into the solution and a transfer of solute from the solution into the food.

Process Variables

Main process variables are

a. Pre-treatments;
b. Temperature;
c. Nature and concentration of the dehydration solutions;
d. Agitation;
e. Additives.

In the light of the published literature, some general rules can be noted:

- Water loss and solid gain are mainly controlled by the raw material characteristics and are certainly influenced by the possible pre-treatments;
- It is usually not worthwhile to use osmotic dehydration for more than a 50% weight reduction because of the decrease in the osmosis rate over time. Water loss mainly occurs during the first 2 hr and the maximum solid gain within 30 min.;
- The rate of mass exchanges increases with temperature but above 45 ° C enzymatic browning and flavour deterioration begin to take place. High temperatures, i.e. over 60° C, modify the tissue characteristics so favouring impregnation phenomena and thus the solid gain;
- The best processing temperature depends on the food; mass exchanges are favoured by using high concentration solutions;
- Phenomena which modify the tissue permeability, such as over-ripeness, pre-treatments with chemicals (SO2), blanching or freezing, favour the solid gain compared to water loss because impregnation phenomena are enhanced;
- The kind of sugars utilised as osmotic substances strongly affects the kinetics of water removal, the solid gain and the equilibrium water content. Low molar mass saccharides (glucose, fructose, sorbitol, etc.) favour the sugar uptake;
- Addition of NaCl to osmotic solutions increases the driving force for drying.

Synergistic effects between sugar and salt have also been observed.

Applications

The effects of osmotic dehydration as a pre-treatment are mainly related to the improvement of some nutritional, organoleptic and functional properties of the product. As osmotic dehydration is effective at ambient temperature, heat damage to colour and flavour is minimised and the high concentration of the sugar surrounding and vegetable pieces prevents discolouration.Furthermore, through the selective enrichment in soluble solids high quality and vegetables are obtained with functional properties "compatible" with different food systems. These effects are obtained with a reduced energy input over traditional drying process. The main energy-consuming step is the reconstitution of the diluted osmotic solution that could be obtained by concentration or by addition of sugar.

Drying

Air drying following osmotic dipping is commonly used in tropical countries for the production of so-called "semi-candied" dried s. The sugar uptake, owing to the protective action of the saccharides, limits or avoids the use of SO2 and increases the stability of pigments during processing and subsequent storage period.

The organoleptic qualities of the end product could also be improved because some of the acids are removed from the during the osmotic bath, so a blander and sweeter product than ordinary dried s is obtained. Owing to weight and volume reduction, loading of the dryer can be increased 2-3 times.

The combination of osmosis with solar drying has been put forward, mainly for tropical. A 24 hour cycle has been suggested combining osmodehydration, performed during the night, with solar drying during the day. Two-three-fold increase in the throughput of typical solar dryers is feasible, while enhancing the nutritional and organoleptic quality of the s.

A two-step drying process, OSMOVAC, for producing low moisture products was described. The osmotic step is performed with sucrose syrup 65-75 Brix until the weight reduction reaches 30-50%. By osmotic dehydration followed by vacuum drying puffy products with a crisp, honeycomb-like texture can be obtained at a cost comparatively lower than freeze-drying.

Commercial feasibility of the process on bananas has been studied, based on the results of a semi-pilot scale operation; the process scheme is reported in Figure 8.4.3. Osmotically dried bananas retained more puffiness and a crisper texture than simple vacuum dried ones, and the flavour lasted longer at ambient temperature. The combination of osmotic dehydration with freeze-drying has been proposed only at laboratory scale.

Appertisation

A combination of osmotic dehydration with appertisation has been proposed to improve canned preserves. The feasibility of a process, called osmo-appertisation, to obtain high quality in syrup, has been assessed on a pilot scale.

The key point of this technique is the pre-concentration of the to about 20-40 Brix, that causes, together with the enhancement of the natural flavour, an increase of the resistance of the to the following heat treatment, especially for colour and texture stability. The products obtained are stable up to 12 months at ambient temperature and show a higher organoleptic quality than canned preserved alternatives. Furthermore, because of their higher specific weight and diminished volume, the filling capacity of jars or pouches is increased.

Freezing

The frozen and vegetable industry uses much energy in order to freeze the large quantity of water present in fresh products. A reduction in moisture content of the material reduces refrigeration load during freezing.

Other advantages of partially concentrating s and vegetables prior to freezing include savings in packaging and distribution costs and achieving higher product quality because of the marked reduction of structural collapse and dripping during thawing.

The products obtained are termed "dehydro-frozen" and the concentration step is generally carried out through conventional air drying, the additional cost of which has to be taken into account. Osmotic dehydration could be used instead of air drying to obtain an energy saving or a quality improvement especially for and vegetable sensitive to air drying.

Extraction of Juices

An osmotic pre-step before juice extraction was reported to give highly aromatic or vegetable juice concentrates.

Further Developments

So far only applications on a pilot plant scale are reported in the literature. For further developments on a larger scale, theoretical and practical problems should be solved.The industrial application of the process faces engineering problems related to the movement of great volumes of concentrated sugar solutions and to equipment for continuous operations. The use of highly concentrated sugar solutions creates two major problems.The syrup's viscosity is so great that agitation is necessary in order to decrease the resistance to the mass transfer on the solution side. The difference in density between the solution (about 1.3 kg/litre) and and vegetables (about 0.8 kg/litre), makes the product float.Another important aspect, so far not investigated, is the microbiological safety of the process, which should be studied thoroughly before further industrial development.

Osmoappertisation in the Processing of Apricots

In order to obtain an alternative to the canned preserves and to maintain a high quality of the s, a research has been carried out on the osmoappertisation of apricots, a "combined" technique that consists in the appertisation of the osmodehydrated apricots.This technique could contributes also to the reduction of energy consumption, limits the cost of production and combines "convenience" (ready-to-eat, medium shelf-life) with many market outlets (retail, catering, bakery, confectionery, semi-finished products).

Osmoappertisation combines two unit operations: dehydration by osmosis and appertisation (packaging + pasteurization).

Apricot Processing

a. Fresh apricot puree After washing, cutting and removal of stones, apricot halves are dipped in 2% solution of sodium or potassium metabisulphite for 10 minutes. After draining, the resulting material is passed through a 0.045-in. screen pulper - finisher to produce a fresh apricot puree.

The fresh apricot puree obtained in this way could be further processed in different semiprocessed (i.e. chemically or otherwise preserved products) or finished products (leathers, bars, jams, etc.).

b. Concentrated apricot pulp Fresh apricot halves could also be steam blanched for 5 min., passed through a 0.045-in pulper - finisher and transformed in a purée with about 14 Brix depending on the quality. This purée may be concentrated in steam jacketed kettles up to 20 Brix or in other adequate equipment (e.g. a stirred vacuum evaporator) up to 28°Brix.

As for fresh apricot purée, the concentrate may be further processed in various semiprocessed or finished products as mentioned above and as will be described below.

c. Dried apricot leather

- From fresh purée by drum drying. The fresh apricot purée at about 14 Brix could be dried to 12% moisture apricot sheets, using a double-drum dryer operating at 132 degrees C with a drum clearance of 0.008 in and speed of 45 sec per revolution.
- From concentrate by drum drying. The concentrate could also be dried to 12% moisture sheets by the same process as described above.
- From fresh purée or from apricot concentrate by sun/solar drying or by dehydration.

a. *Trays:* For sun/solar drying or dehydration of pulp, the trays must have a solid base in order to retain the liquid contents. They may be made of metal, timber or plastic. Stainless steel or plastic trays are most suitable because they are unaffected by acid pulp; they are, however, expensive.

A metal tray could be 75 x 50 cm in size and with side 5 cm high. The trays must keep level during drying; if the tray is not level the pulp will run to the lowest point, giving a layer of irregular depth which will dry unevenly. Any tray which is not made of stainless steel or plastic must be covered inside with a sheet of heavy gauge plastic film to protect the pulp from chemical or bacteriological contamination.

Standard sun/solar trays as described can be used by covering them inside with a sheet of plastic film to create a solid base.

b. *Preparation before drying/dehydration*: Fresh apricot purée can be directly used for next processing steps. concentrate needs to be added to potassium metabisulphite to obtain a 0.3% concentration of SO2 in the material.
c. *Drying/dehydration*: The apricot/ purée or concentrate is poured into the trays to a depth of about 1.5 cm. When stainless steel or plastic trays are used they should be coated with a thin layer of glycerine to prevent sticking.

The pulp is then sun/solar dried or tunnel/cabinet dehydrated; moisture content in the dried product should not exceed 14% and the SO2 content should not be less than 1500 pp. The dried product is wrapped in cellophane to prevent sticking, then put inside polythene bags and stored at best in tight fitting tins and sealed to prevent moisture transfer.

From fresh purée or from apricot concentrate, with sugar addition, and then processed by sun/solar drying or by dehydration. In some countries preference is for finished products with added sugar; and this is also interesting from a point of view of energy consumption (concentration is partially achieved by sugar dry matter) and of shelf life. The overall content in SO2 could also be reduced as sugar is a preservation agent, the product will be close to a "paste".

Reconstitution Test for Dried/Dehydrated Products

In reconstitution water is added to the product which is restored to a condition similar to that when it was fresh. This enables the food product to be cooked as if the person was using fresh or vegetable.

All vegetables are cooked but many of the dried s can be used for eating after they have been soaked in water. The following reconstitution test is used to find out the quality of the dried product.

Reconstitution test

1. Weigh out a sample of 35 grams from the bulked and packed final product of the previous day's production.
2. Put the sample into a small container (beaker) and add 275 ml of cold water (and 3.5 g salt).
3. Cover the container (with a watch-glass) and bring the water to the boil.
4. Boil GENTLY for 30 minutes.

5. Turn out the sample onto a white dish.
6. At least two people should then examine the sample for palatability, toughness, flavour and presence or absence of bad flavours. The testers should record their results independently.
7. The liquid left in the container should be examined for traces of sand/ soil and other foreign matter.

This test can be used also to examine dried products after they have been stored for some time. Evaluation of rehydration ratio may be performed according to the following calculations.

Rehydration ratio. If weight of the dried sample is 10 g (Wd) and the weight of the sample after rehydration is 60 g (Wr), rehydration ratio is:

Rehydration coefficient. The weight of rehydrated sample is 60 g (Wr); the weight of dried sample is 10 g (Wd) and its moisture is 5% (Wu); raw material before drying had 87% water (A); rehydration coefficient is:

= =.

A simpler test for eating quality can be carried out without weighing and measuring. The material is placed in a cooking pot with water (and a little salt). The pot is then covered and boiled as described above.

Except for a few products which are eaten in the dry state, most dried and all dried vegetables are prepared by soaking and cooking. Often this preparation is carried out incorrectly and dried products get a bad reputation.

Good quality dried products, after cooking and if properly treated should be similar to cooked fresh produce. In order to get good results, the following methods are recommended:

Quick Method

Cold water, ten times the weight of the dry product, is added to the dried product. The container is covered, brought to the boil and simmered GENTLY until the product is tender. The cooking time may be 15 to 45 minutes after the boiling point has been reached.

Slow Method

This gives better results than the quick method. Cold water is added to the dry food and is left to soak for 1 to 2 hours before cooking. The product is then cooked in the same water as that in which it was soaked. The actual cooking time will probably be shorter than that for the quick method.

Other points to remember are:

- If too much water is added the cooked product will have little flavour. However, if too little water is added the product may dry and burn. This can be avoided by adding small quantities of water during cooking;
- Always cook with a lid on the container;

- Salt, if required, should be added when the cooking is almost complete;
- Partly used packages of dry products should be reclosed tightly or kept in containers with good fitting lids.

The composition of some dried s is seen in Table given below.

Table. Composition of Some Dried s

	Moisture content %	Sugar as monosac-charides, %	Other carbohydrates	Vitamin C mg/ 100 g
Raisins	21.5	64	7	0.0
Dates	12.5	55	7	0.0
Prunes	20.0	45	15	0.0
Apricots	15.0	45	24	4.8
Peaches	15.5	55	14	8.0

Handling, Sorting, Packing and Storage of Dried and Dehydrated and Vegetables

It is not easy to assess when drying has been completed. In absence of instrumentation, the characteristics of the various products after drying/ dehydration can only be assessed by experience. Although this cannot be conveyed adequately on paper, some general indications can be given.

Products

When a handful of is squeezed tightly together in the hand and then released, the individual pieces should drop apart readily and no moisture be left behind on the hand. It should not be possible to separate the skin by rubbing unpeeled and the centre should no longer reveal any moist area. Banana should be leathery and not too tough to eat in their dry state.

Vegetable Products

Onions should be dried until they are crisp whereas tomatoes should be leathery. In general, the lower the moisture content, the better the keeping quality will be, but overdried products generally have an inferior quality. Also the loss in weight from excessive drying cannot be tolerated in a commercial operation designed to run profitably.

It is, however, essential to dry up to an optimum/safe moisture level, related to the type of the product and his designed shelf life, and to avoid running the risk of the products becoming spoiled due to excess water content. When drying is completed, the material should be sorted either on trays or on a table in order to remove pieces of poor quality and colour and any foreign matter.Very fine material should be separated from the bulk of the material by using a sieve (12 or 16 mesh per inch). Bad quality products which show poor colour need to be removed from the bulk of finished product. After selection and grading, dried products should be packed immediately, preferably in polythene bags which must be folded and closed/tied tightly. However, plastic bags are easily damaged and therefore they must be packed into cartons or jute sacks before they are transported.

Deterioration of Dried During Storage

Dried must be considered as a relatively perishable commodity in the same category as cereals, pulses and similar stored products. It is subject to deterioration resulting from mould growth, insect and mite infestation and physical and chemical changes.

Mould Growth

When the moisture content of dried is allowed to exceed the maximum permissible level for safe storage then mould growth may occur. The moisture levels applicable to various types of; and it can be seen that the safe moisture levels for dried are much higher than those for other similar commodities.

At the present time, suitable field moisture metres for use with dried are not readily available, and moisture determinations can only be satisfactorily carried out where laboratory facilities are available.

Various species of drought resisting fungi may develop on dried when the moisture content is just above the safe level, and a number of osmophilic yeasts are quite commonly associated with spoilage in dried.

Many of the yeasts bring about fermentation with the production of lactic acid or alcohol, and yeasts are frequently present in wart-like crystalline growths which occur in which has become "sugared". In very moist mucoraceous fungi may predominate and are visible as white fluffy growths on and within the.

Mite Infestation

Severe mite infestations are often associated with the growth of osmophilic yeasts in fermenting dried products. Many of these mites are unable to complete their development in the absence of yeast. They have been reported as occurring on dried, and particularly figs and prunes in Mediterranean countries. Such infestations are difficult to eradicate and affect consumer acceptance of the contaminated products.

Insect Infestation

Insect infestations may begin in the field before harvest, may continue during bulk storage after drying, and unless measures are taken to prevent it, may occur in the finished packaged product during storage prior to distribution and consumption.

Regular treatments of the stack of dried with a suitable insecticide will be necessary as a routine to combat light insect infestations. Pyrethrins synergised with piperonyl butoxide are commonly used as a surface spray or as an aerosol fog for this purpose. Heavy infestations will require that the be fumigated.

Equipment for "Dry Sulphuring of before Dehydration/Drying

SO2 generator
1. SO2 outlet pipe
2. Sulphur feeding door
3. Plate for sulphur burning
4. Small door
5. Burner for gas
6. Metallic sieve

Sulphuring installation
1. Exhaust pipe
2. SO2 pipe
3. SO2 generator
4. SO2 distributor
5. SO2 distribution pipes
6. Sulphuring cells
7. SO2 flow regulator
8. Pipe for removing SO2 from sulphuring cells
9. Electrical fan
10. Exhaust pipe for removing SO2 from sulphuring installation
11. Hole for SO2 coming from cells

Solar Dryer with Air Heater - Tunnel Type; Community or Business Level

I. Air heaters
1. Cover: Transparent 3 mm. plexiglass plate or plastic sheet
2. Absorber - bottom - corrugated black painted iron sheets - intermediary - black painted iron sieve
3. Insulation: none
4. Frames: metallic black painted corner/angle ironR

II. Drying tunnel or room
1. Cover: transparent plastic sheets
2. Frames: metallic black painted angle iron
3. Insulation: for back only on the outside
4. Back wall: wood plate, painted black on inside
5. Front wall: transparent plastic sheets
6. Side doors: - frames = wood
   — transparent plastic sheets
7. Dimensions of drying tunnel/room
   — L = 22 feet
   — W= 18 feet

— H = 5 feet

8. One range of trolleys

III. Trays

1. Frames: wood, black painted
2. Bottom: nylon mesh or wood slats
3. Dimensions:
   — l = 3 feet
   — w = 2.4 feet

IV. Trolleys

1. Metal platform with wheels
2. Number of trays per trolley: 15
3. Distance between trays: 3"
4. Total height of trolley: 3 x 15 = 45 inch + 4 " (wheel) = 49 inch

V. Capacity calculation; number of trolleys and trays

1. Hypothesis: 300 kg fresh apricots received per day 2 days solar drying
2. Tunnel/dryer capacity: 2 x 300 = 600 kg fresh apricots
3. Number of trolleys and trays calculation:
   — number of trays/trolley = 15
   — kg fresh apricots per tray = 6 kg
   — kg fresh apricots per trolley: 6 x 15 = 90 kg
   — number of trolleys needed in drying room: 600: 90 = 7
   — number of trolleys in sulphuring cells = 2
   — number of trolleys in "wet preparation section" = 3
   — number of trolleys in dry section (after tunnel) = 2
   — total number of trolleys needed = 14
   — total number of trays needed: 14 x 15 = 200

VI. Sulphuring cell

1. Capacity: 2 trolleys
2. Dimensions: 1 = 5 feet; w = 3.5 feet; H = 55 inch
3. One sulphuring pipe inside the cell, diametre = 15 cm

VII. SO2 generator: diametre = 45 cm; h = 70 cm

VIII. Air heater/heat collector

1. DIMENSIONS: L = 5 M; H = 30 CM; Hl = 15 CM; H2 = 20 CM
2. Bottom/corrugated black painted iron sheets
3. Intermediate heat absorber: iron black painted metallic plate/sieve with holes of about 3 cm in diametre fixation: at mid-distance between cover and bottom of the heat collector
4. Cover: transparent 3 mm Plexiglas plate (or transparent plastic sheets)

The calculation used in this dryer is based on a real situation in a developing country, within the framework of FAO projects; availability of building materials was confirmed. For the orientation of the dryer geographical position of the available drying yard and local wind direction were taken into consideration

Technology of Semi-processed Products

The semi-processed products are manufactured in order to be delivered to industry processing centres (in the producing country itself or in importing countries) where they will be further manufactured in consumer oriented finished products: jams, jellies, syrups, s in syrup, etc.

In the practice of semi-processed products and for the purpose of this document the following categories are defined:

a. "pulps": semi-processed products, not refined, obtained by mechanical treatment (or, less often, by thermal treatment) of followed by their preservation. Either whole, halves or big pieces are used which enables easy identification of the species. "Pulps" can be classified in boiled or non boiled (raw).
b. "purées-marks": semi-processed products obtained by thermal and mechanical treatment or, very rare, raw and then refined, operations by which all nonedible parts (cores, peels, etc.) are removed. "Purées-marks" are classified in boiled (the more usual case) and non boiled (raw).
c. Semi-processed juices: products obtained by cold pressure or very rare by other treatments (diffusion, extraction, etc.) followed by the preservation.

Technical Processes for Preservation of Semi-processed Products

Preservation can be achieved by chemical means, by freezing or by pasteurization. The choice of preservation process for each individual case is a function of the semi-processed product type and the shelf life needed.

Chemical Preservation

In many countries, in practice, this is carried out with sulphur dioxide, sodium benzoate, formic acid and, on a small scale, with sorbic acid and sorbates.Preservation with sulphur dioxide is a widespread process because of its advantages: universal antiseptic action and very economic application. The drawbacks of SO2 are: SO2 turn firms the texture of some species (pomaces), desulphiting is not always complete and recolouring of red s is not always complete after desulphitation.

Practical preservation dosage levels with SO2 for about 12 months is 0.18-0.20% SO2 (with respect to the product to be preserved). This level could be reduced to 0.09% SO2 for 3 months and to 0.12% SO2 for 6 months preservation. The preservation with sulphur dioxide is in use mainly for "pulps" and for "purées-marks".Chemical preservation can be performed from a practical point of view by the utilisation of 6% SO2 water solutions or by direct introduction of sulphur dioxide gas in the product (for "purées-marks"). The preparation of 6% SO2 solutions is done by bubbling the gas from cylinders in cold water; from a 50 kg SO2 compressed gas cylinder results 830 l of 6% SO2 solution.

These SO2 solutions have to be stored in cool places, in closed receptacles and with periodic concentration control/check by titration or by density measurements approximate results.

Preservation with sodium benzoate has the following advantages: it does not firm up the texture and does not modify colour. The disadvantages are: it is not a universal antiseptic, its action needs an acid medium and the removal is partial. Sodium benzoate is in use for "pulps" and for "purées-marks" but less for juices.

Practical dosage level for 12 months preservation is 0.18-0.20 % sodium benzoate, depending on the product to be preserved. Sodium benzoate is used as a solution in warm water; the dissolution water level has to be at maximum 10% reported to semi-processed product weight.

Formic acid preservation is performed mainly for semi-processed juices at a dosage level of 0.2 % pure formic acid (100%). Formic acid is an antiseptic effective against yeasts, does not influence colour of products and is easily removed by boiling.

Formic acid could be diluted with water in order to insure a homogeneous distribution in the product to be preserved; water has to be at maximum 5 % of the product weight. Because of a potential effect of pectic substance degradation, formic acid is less in use for "pulps" and "purées-marks" preservation.Sorbic acid used as potassium sorbate (easily water soluble) can be used for preservation of semi-processed products at a dosage level of 0. 1% maximum. Advantages of sorbates are: they are completely harmless and without any influence on the organoleptic properties of semi-processed products.

Preservation by Pasteurization

As has a low pH, preservation of semiprocessed products could also be performed by pasteurization (heat treatment step at maximum temperature of 100°C), the length of this step varying with the size of the receptacles.The advantages of this type of treatment are: hygienic process, which assure a long term preservation; the disadvantages are: need for air tight receptacles, and pectic substances could begin to deteriorate if the thermal treatment is too long.Thermal preservation of semi-processed products could also be done by a "selfpasteurization": very hot semi-processed products are filled into receptacles (e.g. metal cans) which are sealed and then inverted in order to sterilise the air which goes through the hot mass.

Preservation by freezing

This is done on an industrial scale in some countries and can be done with or without sugar addition. The advantages of this process are: absence of added substances; very good preservation of quality of constituents (pectic substances, vitamins, etc.) and good preservation of organoleptic properties (flavour, taste, colour). Freezing is done at about -20 to -30° C and storage at -10 to -18° C.

Freezing is applied mainly to semi-processed products aimed at very high quality and high cost finished products. Technological flow-sheet for semi-processed "pulps": chemical preservation.

Sorting is needed in order to remove sub-standard (with moulds, with diseases, etc.) and all foreign bodies.

Washing is obligatory in order to remove all impurities which cannot be eliminated at the processing step in finished products.

Coring and Cutting, mainly for pomace s, has as main objective a better utilisation of preservation "space" in receptacles and is not mandatory; this will be defined by customer/supplier agreements/standards. This operation is preferably performed by mechanical means.

Preservation is carried out with the 6% SO2 solution which is added to the prepared s (placed in bulk in receptacles) in the quantity needed to obtain the preservation dosage level. For a better/homogeneous preservative distribution, the initial 6% SO2 solution could be diluted with water; however, the diluted solution (which will be filled in receptacles) has to be at a dosage level of less than 10% of the semi-processed product weight.

For some soft, especially strawberries, preservation is done with a mix of 6% SO2 solution and calcium bisulphite solution (containing also 6% SO2).

Preparation of calcium bisulphite solution is done by the introduction of 30 kg of CaO in 1 m^3 SO2 solution and mixing up to clarification. The resulting solution is mixed with the initial 6% SO2 solution, generally in a 1:1 ratio, but the ratio can be adapted to the fresh texture. Firming of soft texture by this treatment is based on the formation of calcium pectate with pectic substances from tissues.

In the case of sodium benzoate, formic acid or potassium sorbate, the dosage levels to be used are as indicated above with the rule that it is not allowed to add more that 10% liquid in receptacles on the prepared s.

Preservation by pasteurization or "self-pasteurization" will need as additional steps: a) boiling with a minimum water addition (maximum 10%); b) filling of receptacles; c) hermetic closing followed by d) pasteurization or "self-pasteurization".

Some general technical data for the preparation of chemical preserved semi-processed "pulps" are seen in Table given below.

Table. General Technical Processing Data for Semi-processed "Pulps"

species	Preliminary operations	Preservation means
Apples, pears, quinces	Sorting, washing, coring, cutting	Sulphur dioxide

Prunes, peaches, wax cherries, apricots	Sorting, washing, stone removal (pitting)	Sulphur dioxide or sodium benzoate
Cherries	Sorting, washing	Sulphur dioxide or sodium benzoate, sometimes with calcium bisulphite addition
Strawberries	Sorting, washing	Sulphur dioxide in mix with calcium bisulphite
Wild berries	Sorting, washing	Sulphur dioxide or sodium benzoate; in some cases with calcium bisulphite addition

Technological Flow-sheet for Semi-processed "Purée-marks"

The general technological flow-sheet includes the following operations:

Sorting and Washing are obligatory and are carried out in a similar manner as for "pulps".Heat Treatment/Boiling is needed in order to soften the tissues before refining.

For some s as strawberry and wild berries, this step is not done and s are refined "raw" in order to preserve their flavour.Pulping is performed with specific equipment - refiners or pulpers - which eliminate seeds, pits and other non edible parts (peels, cores, etc.).

Preservationis carried out by chemical means, by freezing or by pasteurization.The general technical/processing data for the manufacture of "purée-marks" are seen in Table given below.

Table. General technical Processing Ddata for Semi-processed "Purées - marks"

species	Preliminary operations	Preservation means
Apples, pears, quinces	Sorting, washing, boiling, refining	Sulphur dioxide; in less frequent cases formic acid or sodium benzoate
Prunes, peaches, wax cherries, apricots,	Sorting, washing, boiling, refining	Sulphur dioxide, formic acid or sodium benzoate; cherries freezing with or without sugar; self-pasteurization
Strawberry, wild berries	Sorting, washing, refining	Chemical preservation, freezing or self pasteurization

1. Washing tank
2. Vertical transporter
3. Transport tubes
4. Regulating plates
5. Crushing machine
6. Grating machine
7. Preheating equipment
8. Pulper
9. Centrifugal pump
10. Finisher
11. Storage tank

From the storage tank 11, the product is transferred by pump to a mixing tank (mix with SO2)

Table. Correlation between Density and Concentration for SO2 Solutions

Density at 15°C	Concentration of solutions % SO2
1.0168	3.00
1.0181	3.25
1.0194	3.50
1.0206	3.75
1.0221	4.00
1.0234	4.25
1.0248	4.50
1.0261	4.75
1.0275	5.00
1.0289	5.25
1.0302	5.50
1.0315	5.75
1.0340	6.25
1.0353	6.50
1.0365	6.75
1.0377	7.00

SUGAR PRESERVES TECHNOLOGY; JAMS, JELLIES, MARMALADE, PASTE

As a overall rule of thumb, a sugar concentration of about 60% in finished or processed products generally insures their preservation. Preservation is not only determined by the osmotic pressure of sugar solutions but also by the water activity values in the liquid phase, which can be lowered by sugar addition; and by evaporation down to 0.848 aw; this value however does not protect products from mould and osmophile yeast attack. Maximum saccharose concentration that can be achieved in the liquid phase of the product is 67.89%; however higher total sugar quantities (up to 70-72%) found in products are explained by an increased reducing sugar solubility resulting from saccharose inversion.

Jams

The preservation of by jam making is a familiar process carried out on a small scale by housewives in many parts of the world. Factory jam making has become a highly complex operation, where strict quality control procedures are employed to ensure a uniform product, but the manufacturing operations employed are in essence the same as those employed in the house.Fresh or pre-cooked is boiled with a solution of cane or beet sugar until sufficient water has been evaporated to give a mixture which will set to a gel on cooling and which contains 32-34% water.Gel formation is dependent on the presence in the of the carbohydrate pectin, which at a pH of 3.2 - 3.4 and in the presence of a high concentration of sugar, has the property of forming a viscous semi-solid.During jam boiling, all micro-organisms are destroyed within the product, and if it is filled hot into clean receptacles which are subsequently sealed, and then inverted so that the hot jam contacts the lid surface, spoilage by micro-organisms will not take place during storage. The composition of jam made from stone and berry is shown in Table given below. About 30% of the vitamin C present in fresh is destroyed during the jam-making process, but that which remains in the finished product is stable during storage.

The high moisture content of jam (equivalent to an equilibrium relative humidity of about 82%) makes it susceptible to mould damage once the receptacle has been opened and exposed from some time to the air. No problems of microbiological spoilage are likely to arise in the canned product during storage.

Table. Composition of Some Jams

Type of Jam	Moisture content %	Sugar (as invert sugar, %)	Vitamin C mg/ 100 g
Jam made from berry s: strawberry, raspberry, etc.	29.8	69.0	10 - 25
Jam made from stone s: apricot, peach, etc.	29.6	69.3	10 - 35

Marmalade

This sugar preserve is defined as "semisolid or gel-like product prepared from ingredients together with one or more sweetening ingredients and may contains suitable food acids and food pectins; the ingredients are concentrated by cooking to such a point that the TSS - Total Soluble Solids - of the finished marmalade is not below 65%".

Paste

Paste is a product obtained in the same way as special non-gelified marmalade but with a lower water content - about 25% TSS in paste. Lowering water content could be achieved by continuing boiling of the product or by drying the product by natural or artificial drying.

An example of paste without sugar is the sun dried apricot or prune paste.

General Procedure for the Preparation of Jams, Jellies and Marmalade

a. Boil the pulp or the juice (with water when necessary)
b. Add the pectin
c. Boil for about 2 minutes to assure a complete dissolution
d. Add the sugar while keeping the batch boiling
e. Boil down quickly to desired Brix
f. Add the acid (usually citric acid) and remove the froth

g. Fill hot into the (previously cleaned) jars and close
h. Invert the jars for three minutes to pasteurize the cover

Basic Recipes

The following recipes must be considered only as guidelines because the composition of the can vary; also the taste of the consumers varies concerning the consistency, the sweetness and acidity.

Before starting to make jam it is important to know the yield to settle the question on containers. The calculation is made as follows:

In these basic recipes it is assumed that the s are poor in pectin content.

Recipe 1.: sugar = 50:50; desired Brix in the finished product is 68.

Soluble Solids:

10 kg of at 10% TSS	1.000 kg
10 kg of sugar	10.000 kg
60 g of pectin (grade 200)	0.060 kg
55 g of citric acid	0.055 kg
	11.115 kg

Recipe 2.: sugar = 45:55; desired Brix in the finished product is 68.

Soluble Solids:

d10 kg of at 10% TSS	1.000 kg
2.5 litre of water	-
12.2 kg of sugar	12.200 kg
65 g of pectin grade 200	0.060 kg
60 g of citric acid	0.060 kg
	13.325 kg

Recipe 3.: sugar = 40:60; desired Brix in the finished product is 68.

Soluble Solids:

d10 kg of at 10% TSS	1.000 kg
3.3 litre of water	-
15 kg of sugar	15.000 kg
85 g of pectin grade 200	0.085 kg
80 g of citric acid	0.080 kg
	16.165 kg

Various factors must be taken into account:

1. Size of the container: the quantity of pectin indicated in the recipes is valid for containers of 1 kg or less.

   If container capacity is between:

   Increase pectin by:

1 kg and 2.5 kg	5%
2.5 kg and 5.0 kg	10%
5.0 kg and 10.0 kg	20%
10.0 kg and 20.0 kg	30%

2. Finishing point: the quantity of pectin is given for a final Brix - Total Soluble Solids (TSS) of 68%.

If the final Brix is 66 increase the pectin by 5%

" " is 65 " " " by	10%
is 64 " " " by	15%
" " is 62 " " " by	20%
" " is 60 " " " by	30%

3. Acidic taste. If the product is too acid, replace the citric acid by tartaric acid (63% of the amount of citric acid).
4. Formation of clots: If batch clots, it is probably due to the pH being too low or the or TSS being too high; correct accordingly.
5. Formation of liquid at the surface: if liquid forms on the surface, it is probably due to too low a pH or too low pectin content.
6. Crystallisation:
   a. If liquid forms on the surfaces, then the pH is too low; reduce the acid content;
   b. If liquid does not form on the surface, then TSS or pH is too high.
7. Formation of mould: the TSS is probably below 68 deg. Brix. The filling may have been done at a low temperature. If the containers are large, wait until they are cold before closing.
8. Wrong batch: dilute the jam with water to 30% TSS; cook briefly. Add this diluted jam to a new batch but in a ratio not exceeding 10%.

Processing of Pineapple-papaya Jam

The should be prepared as per previous instructions.

For pineapples, the ends are removed and discarded; the cores and outer parts of the s are also removed. The cylinders obtained are pulped through a special extractor (Fitzpatrick communiting machine) equipped with a 0.40-in screen sieve; the pulp thus obtained is used for making jam.

The papaya are prepared by hand-peeling the; the is then halved and the seeds removed. It is then pulped in the communiting machine using a 0.40-in screen sieve.

When ginger root is used as flavouring, it is peeled and macerated in a Kenwood blender to a very fine consistency.

A typical formula for a pineapple-papaya jam (50:50 ratio) with ginger flavouring is given as follows:

Pineapple pulp	25.0
Papaya pulp	25.0 pounds
Cane sugar	50.0
Apple pectin (150 grade)	6.0 ounces
Citric acid	6.4
Fresh ground ginger	7.5

Processing is carried out in the following way:

The weighed pulp is placed in a stainless steel steam-jacketed kettle and heated to about 110°F under constant stirring.

When the product reaches this temperature, the heat is turned off. The pectin (mixed in about ten times its weight with some of the weighed sugar), is then mixed into the pulp, stirring constantly in order to prevent the pectin from clotting.

When the pectin has dissolved, the remainder of the sugar is added and dissolved completely in the mixture. The heat is then turned on and the jam mixture is stirred constantly until it starts boiling vigorously. During the remainder of the cooking, the product is stirred occasionally. Near the finishing point (approximately 221° F), the citric acid and the ginger (if it is used) are also added.

Determination of the finishing point is done by removing samples at intervals, cooling, and reading the soluble solids by means of a refractometre equipped with a Brix scale. After the jam reaches the proper Total Soluble Solids content, the heat is turned off and the surface scum/foam is removed.

The jam then is quickly put into receptacles which have been cleaned and sterilised with boiling in water for 30 minutes. The filling operation is done rapidly in order to prevent the temperature of the jam from falling below 190° F. After filling, sterilised lids (boiled for 30 minutes in water) are placed on the receptacles and they are then sealed.

After this operation the receptacles are inverted for about 3 minutes to insure that the lids are sterilised. The receptacles are then placed upright. At this stage it is not necessary to do any further processing, therefore the receptacles are cooled in running cold water until they reach a temperature slightly above room temperature. They are then dried in air and labelled.

Evaluation of Finished Products

During production at medium/large scale, it is recommended that quality controls be performed during manufacturing. After ten weeks of storage at room temperature it is recommended that an examination of finished products be performed. The receptacles are opened and contents carefully emptied on to enamel trays without disturbing the formation of the jam.

The empty cans (if metal cans were used) are then inspected for signs of corrosion. Factors other than flavour include colour, appearance, syrup separation, firmness and spreading quality. For flavour, jam is tested on pieces of bread. Samples are taken for measurement of pH (with a glass electrode pH metre) and Total Soluble Solids (with a refractometre equipped with a Brix scale).This evaluation enables to have a quality check during product shelf life and to obtain data needed for necessary improvements of future productions.

For pineapple-papaya jam, products made with 30% pineapple and 70% papaya with added ginger has the highest score for flavour. The use of plain tin cans causes corrosion problems which is not the case when acid resistant lacquer cans are used.

Pineapple Jam Making

1. Boil 40 lb. of pulp and 12 lb. of water.
2. Add 225 g of pectin to the batch while stirring rapidly.
3. Boil for about 90 sec to assure complete dissolution.
4. Add 60 lb. of sugar gradually if possible in several portions, while keeping the batch boiling.
5. Boil down quickly to 69 deg. Brix (223 deg. F).
6. Take off steam, remove foam.
7. Add 300 cc citric acid solution 50%.
8. Fill hot (180 deg. F).
9. Invert receptacles for 3 minutes.

Check each batch for Brix 68-70 deg.; acidity/pH = 3.2 +/- 0.2.

Evaluate: absence of defects; colour; flavour; consistency.

Various:sugar ratios can be manufactured; some basic recipes are as follows:

	Ratio 50:50	Ratio 45:55	Ratio 40:60
	55 lb	49.5 lb	44 lb
Water	-	11 lb	13.2 lb
Sugar	55 lb	60.5 lb	66 lb
Pectin (150 grade)	225 g	237.5 g	250 g
Citric acid (50% sol.)	300 cc	320 cc	335 cc

Gelified Sugar Preserves

Technology of Jellies

Jellies are gelified products obtained by boiling juices with sugar, with or without the addition of pectin and food acids. Jellies are usually manufactured from juices obtained from a single species only, obtained by boiling in order to extract as much soluble pectin as possible.

Jellies have to be clear, shiny, transparent and with a colour specific to the from which they are obtained. Once the product is removed from the glass receptacles where it was packed, jellies must keep their shape and gelification and not flow, without being sticky or of a too hard consistency.Technological flow-sheet for jellies manufacturing covers two categories of operations: those to obtain gelifying juices and those related to the manufacturing of jelly itself.

a. Production of gelifying juices: Washing & Sorting are carried out in usual conditions; Cutting is applied eventually only to pomaces (apples, quinces) and are limited to cutting in halves or quarters;
   Boiling is performed with water addition with 50-100% water, needed for pectin extraction. The boiling time is 30-60 min., it should not be longer so as to avoid pectin degradation; at the same time boiling must not be too violent.

Juice separation is carried out by a simple drain through metallic sieve or cloths; in these cases the yield is lower and the residue can be used for marmalade production. In bigger productions it juice separation by hydraulic press is preferred, yield being in these cases greater.

Juice clarification is strictly necessary in order to obtain clear jellies. This step can be achieved by sedimentation during 24 hours or by filtration.

b. Manufacturing of jellies:

Basic Recipe Seting is done starting with equal parts in weight of sugar and juice (for example 1000 g juice and 1000 g sugar). As final jelly has to contain about 60% added sugar, weight of finished product must be of about 1600 g, by evaporation of about 400 g water.

Boiling is carried out as following: juice is boiled up to removal of about half of the water that has to be evaporated, then the calculated sugar quantity is added gradually; the remainder of the water is evaporated until a concentration in soluble substances (refractometric extract) of 65-67% is reached, in which is incorporated also the sugar from juice.

During boiling it is necessary to remove foam/scum formed. Product acidity must be brought to about 1% (malic acid) corresponding to pH > 3. Any acid addition is performed always at the end of boiling.

For juices rich in pectin, gelification will occur without pectin addition. If at the trial boiling test the gelification has not occurred, because of pectin absence, in this case 1-2% powder pectin will be added by operating as indicated: pectin is mixed with 10-20 fold sugar quantity and is introduced directly in the partially evaporated juice and then boiling is conducted rapidly up to final point. Evaluation of final point is done not only by refractometry but also by gelifying test.

A rapid test for evaluation of juice pectin content is possible by mixing a small sample of juice with an equal volume of 96% alcohol; the apparition of a compact gelatinous precipitate indicates a sufficient pectin content for gelification. Boiling of jellies is performed in small batches (25-75 kg) in order to avoid excessively long boiling time which brings about pectin degradation. Cooling is optional and is carried out up to 85 deg. C, in double wall baths with water circulation. Filling is performed at a temperature not below 85 deg. C in receptacles (glass jars, etc.), which must be maintained still about 24 hours to allow cooling and product gelification.

Receptacle closing is done after product gelification.

Usual jelly types are: quinces, strawberries, cherries, wild berries, alone or in mixes with apples.

Grading of Marmalades

Three categories can be defined:

- Fine marmalade, manufactured from one;
- Superior marmalade, obtained from a mix of in which 30% are "noble" species (cherries, strawberries, apricots, etc.) and 70% from other species;
- Marmalade from mixes; apples, pears, plums, quinces, ungrafted apricots and wax cherries may be used, with the optional addition of "superior" which was rejected at sorting but which was sound.

The content in total soluble substances (refractometric extract) of marmalades must be 64% minimum; the acidity must be between 0.5% and 1.8% expressed as malic acid.

Basic Recipe Setting For a normal composition - marmalade without pectin addition the following is a basic recipe:

100 kg semi-processed product
(10% refractometric extract) .... 10 kg soluble substances
55 kg sugar .... 55 kg soluble substances
155 kg 65 kg soluble substances
55 kg water to be evaporated
100 kg marmalade with 65% refractometric extract

This marmalade satisfies many standards and at same time has a good shelf-life since it contains less than 35% water.

Semi-processed products must have a minimum 8% refractometric extract; in this case the recipe should use 125 kg of raw material, with 80 kg water to be evaporate.

The use of semi-processed products with a low refractometric extract presents the following drawbacks:

a. Higher water quantity to be evaporated;
b. Longer boiling times with negative impact of pectin degradation;
c. Loss of flavour and
d. Lower equipment efficiency.

Pectin addition in marmalade manufacture produces the following advantages:

a. Improvement of gelification,
b. Economy in;
c. Shorter boiling time; this maintains taste and flavour and produces higher equipment efficiency.

Pectin addition makes it possible to obtain the "fine" type of marmalade from "noble" s which do not contain enough pectin (cherries, peaches, apricots, etc.). In marmalades from mixes, low pectin content can be compensated by addition of semi-processed products which are rich in this component (for example apples).

When pectin is to be added, the above recipe should be modified as follows:

80 kg semi-processed product
(10% refractometric extract) .... 8 kg soluble substances
55 kg sugar .... 55 kg soluble substances
10 kg pectic extract (10 % R.E.) .... 1 kg soluble substances
145 kg 64 kg soluble substances
45 kg water to be evaporated
100 kg marmalade with 64% refractometric extract

Pectin can be added as pectic extract with about 10% refractometric extract (R.E.) in the recommended proportion or in the form of a powder considered with 100% dry matter (e.g. 100% soluble substances) in a quantity of about 1%.

In some countries it is usual to add 5-12% of corn syrup (calculated to finished product weight) to replace the corresponding quantity of sugar (100 parts corn syrup can replace 80 parts sugar). Corn syrup has to be liquefied by heating before use. Corn syrup addition reduces the too sweet taste of marmalade, avoids sugar crystallisation and gives a special shine to finished products.Marmalade manufacturing follows the technological line drawing in Fig. 8.6.1 and covers the following steps:

"MARK" Preparation can be achieved from fresh s (as indicated in the section related to semi-processed products) or starting from chemical preserved semiprocessed products: "marks" or "pulps". In the latter case, pulps will be processed in marks which then will be desulphitated.

Desulphitation is carried out by boiling at atmospheric pressure, under vacuum in specialised equipment or under pressure in special retorts built in acid-resistant material. In any case, the desulphitation must be carried out before sugar addition because sugar will bound to the sulphur dioxide. The desulphitation operation must be conducted so as to be, if possible, fully completed; the finished product must contain less than 0.005% free SO2.

The technological flow for marmalade production is the following: fresh after sorting on control belt (1) is washed in a washing machine (2) is brought to the continuous boiling equipment (3) then to the pulper (4); the semi-finished mark is passed on to the storage tank (6). Pulps are boiled and desulphited in continuous boiling equipment (3), then are brought to pulper (4) and to the storage tank (6).

Boiling aims at evaporating the required water quantity, to facilitate the formation of pectin-sugar-acid gel and to partially invert sugar (about 40% from total sugar). The boiling operation can be carried out in open kettles or in evaporators under vacuum.

In the latter case, the warm "mark" from storage tank (6) is aspirated in a concentrator (7) in a vacuum and submitted to a partial boiling up to removal of half of water quantity which needs to be evaporated; the calculated sugar quantity is then added by aspiration, keeping the boiling on. After this the pectin extract or powder pectin which has previously been dissolved in warm water, is added; when the final concentration is reached, as indicated by refractometric control, the required quantity of acid is added. Sugar is added in proportion of 55% in finished product, pectic extract (10% refractometric extract) at a level of about 10-15% and the acid (citric, tartaric, lactic) in a quantity needed to obtain a finished product acidity of about 1%.

Boiling at atmospheric pressure affects not only the appearance but also the nutritional value of the products, mainly if these contain proteins, as some albuminoids coagulate even at 60°C.

Food products for which flavour is an essential property as for example juices, etc., are also affected by the action of heat. Heat treatment has an impact on vitamin losses, mainly of vitamin C, in the presence of oxygen as is the case at concentration in open vessels.Sugars are generally less damaged by heat at temperatures below 100° C; as the boiling point is increasing above 100° C, a risk of partial sugar caramelization exists. In the study of heat effects on products submitted to concentration operation, it is necessary to take into account not only the evaporation surface temperature but also the distribution of the temperature in the whole liquid mass.

The length of the heating period also has a major influence because in many cases it is preferable to concentrate the liquid at a relatively high temperature in a short time avoiding the drawbacks of lower temperatures acting during a long time.In order to maintain the food value and organoleptic properties, it is necessary that concentration take place at a low temperature which can be achieved by concentration under partial vacuum, taking into account that boiling point decreases when the residual pressure decreases, respectively with the increase of vacuum degree.

Advantage of concentration under partial vacuum are the following:

- Lowering of boiling point;
- The total time needed for concentration of food products under a residual pressure of about 200 mm Hg is about half as compared to the that of concentration by boiling at atmospheric pressure;
- By lowering the concentration temperature and time, organoleptic properties and of nutritional value are maintained better particularly as far as the vitamins are concerned;
- When products are concentrated in a vacuum, it is possible to recover volatile aromatic substances by using adequate installations.

Technical procedures of concentration by vaporisation can be classified in:

a. Concentration at atmospheric pressure: continuous or discontinuous;
b. Concentration under partial vacuum: discontinuous (in vacuum

equipment with simple or multiple effect) or continuous (in vacuum installations with continuous action or in thin film vaporisation installations).

Even if open kettle equipment is less expensive than evaporators in a vacuum, it is necessary to take into account that boiling under vacuum has the following advantages:

a. Low boiling temperature (60-70° C), depending the degree of the vacuum; this give the better taste and flavour-keeping qualities;
b. Easy feeding with raw and auxiliary materials;
c. Shorter boiling time;
d. Better working conditions (vapour elimination in condensed water and not in open air).

There are small size evaporators under vacuum which can be well suited to the needs of medium size operations in developing countries.

Cooling of marmalade to about 50-60°C can even be done in a vacuum evaporator by closing the heating steam and maintaining vacuum degree or by discharge in storage tanks (8).

Filling in receptacles (boxes, jars, glasses, etc.) is done preferably with filling machines (9) followed by labelling(10). Small packages can be closed warm or after complete cooling; big packages (boxes, etc.) must be closed only after cooling, e.g. 24 hours after processing.

STORAGE of marmalade must be done in dry rooms (air relative humidity at about 75%), well ventilated, medium cool places (temperature 10-20 degrees C), disinfected and away from direct sunlight and heat. These measures are necessary because marmalade is a hydroscopic product and, by water absorption, favourable conditions for mould development are created.

Technology of Special Jams

Special jams are products similar to marmalades but in which partially keep their shape (whole, halves, etc.). Special jams are obtained by boiling with sugar, with or without pectin addition, with acid addition followed by n concentration by evaporation.Special jams present a pronounced gelification at their cooling and can be considered as s included in a pectin-sugar-acid gel.High quality special jams are obtained only from fresh or possibly frozen and from only one species.

Special jams are classified in:

a. Non-pasteurized (min. 68% refractometric extract) and
b. Pasteurized (min. 65% refractometric extract); minimum acidity, expressed in malic acid, is 0.5%.

Basic Recipe Seting is done taking into account following elements: - maintenance, as much as possible, of shape, because this is specific to these finished products;

- Added sugar, in relation to finished product, must be at 60-65%; this high percentage is needed in order to obtain a rapid and strong gelification and to facilitate conservation of shape;
- A satisfactory gelification cannot generally be obtained without pectin addition, at a level of 1-2%, or respectively 10-20% pectic extract (10% refractometric extract);
- Partial (about 40%) inversion of added sugar is necessary.

The basic recipe is: 80 kg + 100 kg sugar + 1.6-2.0 kg pectin powder and 1 kg citric or tartric acid; this will yield about 165 kg of special jam with 60% added sugar; water quantity to be evaporated is about 18 kg. If the finished product has to contain 65 % added sugar, the boiling has to be continued up to evaporation of about 30 kg water, the resulting finished product quantity will be about 153 kg.

Technological flow sheet for manufacturing of special jams is as follows:

Preparation: sorting, washing, peeling and coring (for apples, pears, quinces), or removal of quetches and stones/pits (for plums, peaches, apricots, cherries) or of short tails (for strawberries and wild berries). Pomace s are then cut in slices or quarters.

Boiling with sugar is the most important operation in production of special jams and has as objective to evaporate water until gel formation and partial inversion of sugar.

Boiling has to be conducted in such a way as to avoid disintegration, but s must be well penetrated with sugar. By boiling an equilibrium is reached, by osmosis, between sugar solution and cellular juice. The initial concentration gap between sugar syrup and cellular juice is very high and if the equilibration process is forced, juice comes out of cells and loses its shape and may even disintegrate.

The boiling process accelerates the equilibration, intensity of which increases with temperature and boiling time.

Pectin addition shortens boiling and thus delays the equilibration; for this reason there are different methods for special jams preparation:

- Maintaining in sugar, at ambient temperature, over 8-24 hours; to sugar ratio is that indicated in recipe. After this sugar impregnation, s and resulting sugar syrup are brought together to boil; toward the end of boiling, pectin, dissolved in warm water, and necessary acid quantity are added.
- Preparation of a very concentrated sugar syrup (at least 75%), in which is introduced in order to be boiled. The rest of the operations are similar to those described in the previous method.
  - Soft (strawberries, wild berries) can be mixed with sugar directly in evaporating open kettles, without added water and then heated gradually up to boiling, which is continued as in previous method.

Boiling is preferably carried out in small open kettles (50-100 kg) in order to avoid too long a boiling and disintegration.

Gelification corresponds generally to the reaching of a concentration of 65% soluble extract, respectively 68% refractometric extract. Practical test for gelification is done as for jellies and marmalades.

Cooling is a technological step strictly necessary in order to avoid rising to the surface in preservation receptacles. Cooling is done in a double bottomed water bath in which water circulates at about 80° C.

Filling of receptacles (jars, boxes, glasses, etc.) is carried out and it is necessary to assure at this stage that the finished product is homogeneous (equal quantities of s and gel).Pasteurization is only applied to special jams with 65 % refractometric extract packed in jars or boxes and is performed at about 100° C for about 20 min.

Gelification is carried out during product cooling and intensifies during storage.

Storage must follow the same conditions as for marmalades.

Non Eelified Sugar Preserves

Technology of Jams

Jams are products obtained by boiling of s (whole, halves, etc.) with sugar syrup until the reaching a viscous consistency. Jams can be defined as s included in a concentrated syrup.Jams are only prepared from fresh s; the usual product range covers the following species: strawberries, cherries, apricots, wild berries, peaches, plums, raisins, quinces, rose petals, etc.; manufacturing of jams from mixes is not an accepted practice. Standards indicate that jams have to conform with following conditions:

- or pieces content 45-55%;
- Soluble substances, refractometric degree min. 72%;
- Acidity (expressed in malic acid) min. 0.7%.

Basic Recipe Setting is presented below as an indication:

100 kg s
150 kg sugar
250 kg
35 kg water to be evaporated
215 kg jam at about 72% refractometric extract

Preparation is similar to special jams with the difference that stone/seed removal is mandatory for all species. Boiling s with Sugar can be achieved by the same three methods described above for special jams. By boiling a sugar content equilibration is foreseen and the operation must be conducted in such a way that texture, flavour and colour of s be preserved. Foam/scum has to be removed during boiling.

Generally boiling in concentrated sugar syrup (at least 75%) after a previous diffusion during 2-4 hours is in practical use. Boiling must to be carried out in small portions (about 15 kg) in order to avoid disintegration.

For some species, boiling has to be conducted in many phases/steps with "stops" to enable sugar diffusion. At the end of boiling, vanillin at a ratio of 125 g for 100 kg jam may be added for some s (white cherries, raisins, etc.). At the same time, it is also possible to add citric or tartric acid in order to avoid the "sugaring" defect.

Cooling of jams, necessary in order to avoid rising to the surface, is carried out as for special jams.Filling of receptacles and Storage for jams are performed in same way as for special jams.

Technology of Special non Gelified Marmalades

Special non gelified marmalades are products resulting from without stones or seeds, sieved or squashed, concentrated by boiling, without sugar added and non gelified. Their consistency results from a low water content (about 35%) and a high percentage of insoluble substances (5-10%). Sugar the from s acts as a preserving agent.

Plum special non gelified marmalade is the product representative of this category. Other is used very rarely, because they have a reduced sugar content as compared to plums; though there are some countries producing special non gelified marmalade from pears or sweet apples.

For plums, the finished product in this category must contain minimum 55-60% soluble substances (refractometric extract), rising up to 70% for a high quality product.Basoc Reco[e Settomg for plums is done in relation with sugar content of s; when this is higher, the product quality is better and the yield is higher. Thus, if plums with 18% refractometric extract are used, 300-350 kg of s are needed for 100 kg finished product.

At the same time, concentration of sugar by boiling, also increases by three fold acidity and astringency of this special plum non gelified marmalade; for this reason it is recommended to use only sweet and completely mature plums.Washing is performed in usual conditions.

Pre-Boiling of s can be carried out in water or vapour, preferably with continuous running and has as its objective the softening of tissues.

De-Stoning is performed in a pulper. Boiling of the sieved mass is done in double bottom open kettles with a big evaporation surface or in vacuum evaporators. Boiling in open kettles enables production of a more tasty slightly caramelized, product; boiling in vacuum evaporators has the technological advantages indicated in marmalade production.

At the end of boiling and once of necessary concentration is reached, the product is poured directly into receptacles (drums, etc.) and left to cool in order to form a hard surface layer (crust).

Storage of well closed receptacles is carried out as for marmalade.

Special non gelified marmalade can also be prepared from chemically preserved semi-processed products, but the quality is lower than that obtained from fresh s.Sometimes dried prunes in a mix with preserved semi-processed products can be used for plum special non gelified marmalade preparation.

In some countries plum finished products in this category are sweetened by the addition of maximum 30% sugar, calculated in relation to the finished product.

Technology of Pastes

These products are obtained in a similar way to marmalades and special non gelified marmalades, but have a lower water content (about 25%). Reduction of water content can be achieved by continuing the boiling of the product or by natural or artificial drying.

A typical example of paste without sugar added is the apricot paste - "pistil", etc. which is a concentrated special non gelified marmalade poured in thin layers and sun dried. An example of paste with sugar added is quince paste which is a marmalade concentrated by evaporation. Sugar content must be 65%; soluble substances content, 7075 % refractometric extract and acidity at least 0.5% expressed as malic acid. Packing is done usually in polyethylene sheets and then in boxes or tins; storage conditions are similar to those for marmalade.

Technology of Syrups

syrups are products obtained by dissolving sugar in juices obtained from direct pressing of s. Sugar dissolving can be done at room temperature or by heating.Syrups have to contain 68% soluble substances (refractometric extract) and minimum 1 g/100 ml malic acid. Up to a maximum 10% of sugar can be replaced by corn syrup. Syrups must be manufactured from the juice of only one species.

Juice preparation is carried out at room temperature as described above.

Sugar dissolving in juice can be performed by one of the following methods:

a. Boiling in open kettles, done by using the basic recipe:
   350-400 kg juice; 650-660 kg sugar; max. 10 kg citric or tartric acid. The juice is brought to boiling and the sugar is dissolved; the total time has to be as short as possible in order to avoid flavour loss and a too high sugar inversion degree (optimum inversion degree is 40%). Acid is added preferably towards the end of boiling. During all boiling processes it is necessary to remove foam/scum. In order to avoid caramelization, the syrup has to be cooled rapidly, and this can be carried out in baths with double bottoms through which are circulated water. One alternative to this method is to boil syrup in closed vessels to avoid flavour losses.

b. Boiling in a vacuum. The basic recipe is the same as above. Sugar and juice are mixed previously in a pre-heating kettle and then transported to vacuum equipment.
   Boiling is performed at 50°C and at the end the temperature is raised slowly up to 65-70°C. The syrup can be cooled directly in vacuum equipment by closing the steam inlet and by increasing the vacuum. In this boiling method it is possible to incorporate a flavour recuperation device.
c. A continuous process for syrup preparation can be carried out by dissolving components with heat while passing them through a horizontal cylinder with a screw inside.
   In the methods where sugar is dissolved by heat, it is also possible to use chemically preserved juices. In this case it is necessary to first perform the desulphitation of juices preserved with SO2. This can be performed by boiling juice with optional water addition (and before any sugar addition). High quality syrups are obtained however from fresh juices.
d. Sugar can also be dissolved at room temperature by using continuous flow percolators. These are similar to those used for salt solution preparation in vegetable canning processes. The juice goes over a sugar layer and is concentrated progressively until saturation (about 65 %). The syrup is then passed through a filtration section in the bottom of the percolator.
   Syrup filtration is needed in order to clarify crystals; the filtration of syrup is done in warm conditions through cloth.
   Filling of syrup in bottles is done in aseptic conditions as much as possible in order to avoid syrup infection with osmophile yeasts.
   Syrup preservation is assured by the high sugar content with respect to a low water activity [unclear].
   Storage takes place in well ventilated storage rooms; avoiding sunlight at 10-15°C.
   The usual product range is: strawberries, cherries, wild berries, citrus s.

Technology of " in Syrup" Products

This type of product is represented by (whole, halves or pieces) covered by a sugar solution and preserved by pasteurization. In these products sugar does not have a preservation effect but only a sweetening role.

Technological general flow-sheet covers the following steps:

Sorting is necessary in order to choose mature, whole, unblemished, undamaged and with a specific texture. of good quality but with a texture not compatible with this type of finished product are used for the production of semi-processed products, marmalades, etc. This step is done manually on a sorting belt.Washing is performed in equipment with fan and sprays. Cleaning covers removal of leaves, etc., skin removal by one of the described methods for some s and coring/pitting for others.

Cutting is applied only to pomace s and is performed preferably by mechanical means.Preliminary Heat Treatment depends very much upon the types and can be as different as a light blanching up to a real boiling; this step is aimed toward softening of tissues for hard s, elimination of waxy pruin layer (plums, etc.) or enzyme inactivation for pomace s.

This treatment has to be reduced to the minimum necessary in order to avoid sugar losses. Some (for example apricots, black cherries, grapes, etc.) does not undergo preliminary heat treatment.Cooling is carried out in water (as cold as possible) and should not be too long to avoid soluble substances loss.

Receptacle filling is introduced manually or sometimes mechanically (filling tables, etc.) in receptacles and then sugar syrup is added.

Sugar syrup preparation is performed by dissolving crystal sugar in hot water (90100° C). After the sugar has been dissolved, the syrup is boiled briefly; removal of impurities and coagulated substances is then performed by foaming/ scumming. An addition of about 0.3% citric acid helps syrup clarification, followed by filtration through cloth.

Preheating/Emptying of open receptacles has as its objective to eliminate air from tissues and this step is largely replaced today by receptacle closing under vacuum. Preheating may be done in steam or in water in such a way that syrup temperature reaches 8090°C and is maintained for 10-15 min.

Hermetic sealing is performed according to the type of receptacle and of cover/cap.Pasteurization is carried out at 100°C and may be performed in continuous installations or in normal water baths.

In some countries s in syrup are pasteurized at 80-90°C; this contributes to a greater flavour retention.Cooling should be intensive in order to avoid discolourations and colour modifications and to reduce corrosion in metal cans.

Refractometric extract is to be measured 30 days after manufacturing. In order to establish exact concentration of sugar syrup to be added in each individual case it is possible to use the following formula:

$C = 1.5 \times [(2 \times E) - F]$

C = added syrup concentration;

E = refractometric extract needed in finished products, according to Table 8.6.3.

F = refractometric extract.

Hot syrup is put into receptacles with a minimum temperature of 80°C in order to enable a corresponding vacuum after cooling.

It is possible to add in the syrup 0.3-0.5% citric or tartaric acid for with a low acidity. Adding ascorbic acid in proportion of 0.8% assures colour maintenance and taste improvement based on its reducing action.

Avoidance of excessively soft texture in finished products may be achieved for soft (strawberries, apricots, etc.) by dipping in a solution of calcium chloride (5% $CaCl_2$ in water).

Finished product defects and production "accidents" for " in syrup" and the means of preventing them

- Change of colour in red (red cherries, plums, etc.) to violet; to avoid this it is necessary to use glass jars or varnished/lacquered tinplate;
- Colour change to red for pears and quinces (pink shade) could arise in products which are over-pasteurized and insufficiently cooled;
- Change of colour of whole peaches peeled by chemical means is due to an incomplete inactivation of oxidative enzymes; to avoid this it is necessary to boil after chemical peeling followed by washing with 0.5% citric or tartric acid solutions or by adding ascorbic acid in syrup;
- Softening of strawberries tissues can be prevented by a pre-treatment with warm syrup and also by addition of calcium salt.

Some technical data for processing are seen in Table given below.

Table. Technical Data for " s in Syrup"

Type of/ product (in syrup), name	Size/shape	Minimum RE** at content, %	Soluble substances 20°C, minimum
Apricots	Halves	55	25
Strawberries	Whole	47	25
Cherries	Whole	53	22
Quinces	Slices, 30 mm	50	26
Apples	Halves or quarters	47	25
Pears	Halves or quarters	57	22
Plums	Whole	45	25
Melon	Pieces	55	26
Wax cherries	Whole	53	27
Wild cherries	Whole	47	27
Pineapple	Slices	55	26
Mangos	Pieces	50	25
Papaya	Pieces	55	27
mixes	According to the conditions indicated above, for each	50	25

Juice Technologies

juices are products for direct consumption and are obtained by the extraction of cellular juice from, this operation can be done by pressing or by diffusion.

For the purpose of this document, the technology of juice processing will cover two finished product categories:

- juices without pulp ("clarified" or "not clarified");
- juices with pulp ("nectars").

We will also define as

a. "natural juices" products obtained from one; and
b. "mixed juices" products obtained from the mix of two or three juices from different species or by adding sugar.

Juices obtained by removal of a major part of their water content by vacuum evaporation or fractional freezing will be defined as "concentrated juices".

Technological Steps for Processing of Juices Without Pulp

juices must be prepared from sound, mature only. Soft varieties such as grapes, tomatoes and peaches should only be transported in clean boxes which are free from mould and bits of rotten.

Washing: must be thoroughly washed. Generally, will be submitted to a pre-washing before sorting and a washing step just after sorting.

Sorting: removal of partially or completely decayed is the most important operation in the preparation of for production of first quality juices; sorting is carried out on moving inspection belts or sorting tables.

Crushing/Grinding/Disinteration Step is applied in different ways and depends on types: Crushing for grapes and berries; Grinding for apples, pears; Disintegration for tomatoes, peaches, mangoes, apricots.

This processing step will need specific equipment which differs from one type of operation to another. Enzyme Treatment of crushed mass is applied to some s by adding 2-8% pectolitic enzymes at about 50° C for 30 minutes.

This optional step has the following advantages: extraction yield will be improved, the juice colour is better fixed and finished product taste is improved. However, for which is naturally rich in pectic substances, this treatment makes the resulting "exhausted" material useless for industrial pectin production.

Heating of crushed mass before juice extraction is an optional step used for some in order to facilitate pressing and colour fixing; at same time, protein coagulation takes place. Pressing to extract juice.

Diffusion is an alternative step for juice extraction and can be carried out discontinuously or in batteries at water temperature of about 80-85 ° C.

Juice clarifying can be performed by centrifugation or by enzyme treatment. Centrifugation achieves a separation of particles in suspension in the juice and can be considered as a pre-clarifying step. This operation is carried out in centrifugal separators with a speed of 6000 to 6500 RPM. Enzyme clarifying is based on pectic substance hydrolysis; this will decrease the juices' viscosity and facilitate their filtration. The treatment is the addition of pectolitic enzyme preparations in a quantity of 0.5 to 2 g/l and will last 2 to 6 hours at room temperature, or less than 2 hours at 50° C, a temperature that must not be exceeded. The control of this operation is done by checking the decrease in juice viscosity. Sometimes, the enzyme clarifying is completed with the step called "sticking" by the addition of 5-8 g/hl of food grade gelatine which generates a flocculation of particles in suspension by the action of tannins.

Filtration of clarified juice can be carried out with kieselgur and bentonite as filtration additive in press-filters (equipment). De-Tartarisationis applied only to raisin juice and is aimed to eliminate potassium bitartrate from solution. This step can be performed by the addition of 1% calcium lactate or 0.4% calcium carbonate. Pasteurization of juice can be done for temporary preservation (pre-pasteurization) and in this case this operation is carried out with continuous equipment (heat exchangers, etc.); warm juice is stored in drums or large size receptacles (20-30 kg). Pasteurization conditions are at 75°C in continuous stream. Pasteurization of bottled juice is then carried out just before delivery to the market; this is performed in water baths at 75° C until the point where the juice reaches 68° C. In cases when the final pasteurization is done without pre-pasteurization and temporary storage, modern methods use a rapid pasteurization followed by aseptic filling in receptacles.

Rapid pasteurization conditions are as follows: temperature about 80° C, over 10-60 sec., followed by cooling; all operations are carried out in continuous stream. Preservation under CO2 pressure may be done at a concentration of 1.5% CO2 under a pressure of 7 kg/cm^2. At the distribution step, proceed at CO2 decompression and the juice is then submitted to a sterilising filtration and aseptic filling in receptacles. Preservation by freezing is carried out at about -30° C, after a preliminary de-aeration; storage is at -15 to -20° C.

Production of concentrated juices by evaporation is performed under vacuum (less than 100 mm Hg residual pressure) up to a concentration of 65-70% total sugar which assures preservation without further pasteurization. Modern evaporation installations recover flavours from juices which are then reincorporated in concentrated juices.

Additional operations for juice manufacturing are the vacuum de-aeration and mixing with other juices or with sugar. For the production of non clarified juices the centrifugation is the only specific step, enzyme clarifying and subsequent filtration being eliminated. The optimum sugar/acid ratio for the majority of, mainly for pomaces, is 10/1 to 15/1. which is rich in carotenoids (apricots, peaches, etc.) is only processed as juices with pulp ("nectars").

Technological Flow-sheet for Juices with Pulp ("Nectars")

This process is divided at industrial scale in two categories of operations:

a. Processing for obtaining juices;
b. Juice conditioning for preservation.

a. Operations in the first category are differ according to the type of which to be processed.

Pomaces (apples, pears) are washed and sorted and then crushed in a colloid mill; purée is then passed through a screw type heating equipment where direct steam is used as a source of heat. Warm mass is treated in a pulper with a 2 mm screen and then through an extractor similar with the equipment used for tomato juice. Stone s (apricots, peaches, cherries, etc.) after washing and sorting are submitted to steam in a continuous heater, then the warm mass is passed through a pulper and then an extractor (as mentioned above). Berries (strawberry, wild berries, etc.) are washed, sorted and then crushed, preheated and then introduced in extractor.

In order to avoid browning and undesirable taste modifications it is usual to add about 0.05% ascorbic acid.

b. Second category type of operations are similar for all species. Partial elimination of cellulose is achieved with a continuous centrifugal separator; the resulting juice is then processed in order to adjust sugar and acid content for viscosity.

Sugar (about 8-10%) is added as a syrup (in water or in the juice of same obtained by pressure). Acidity is adjusted with citric or tartaric acid. The adjusted juice is then deaerated under vacuum at about 40° C. This step aims at avoiding oxidative reactions and vitamin C loss reduction. An important subsequent step is an intensive homogenisation (under pressure at 150-180 A) in order to obtain particles with dimensions below 100. The homogenised juice obtained is then continuously pasteurized in plate heat exchanger equipment at a temperature of about 130° C, cooled down to about 90° C and aseptically packed in receptacles.

The principal characteristics of "nectars" are uniformity and stability of the content provided by the advanced disintegration of s. Stability can be obtained by increasing product viscosity by adding pectin for which is deficient in this component. In order to avoid "separation", intensive homogenisation is carried out as described above. "nectars" contain all the important components of the original and to a large extent maintain their taste and flavour. The sugar/acidity (as citric acid) ratio is to a large extent determined by the type of and the correction applied; for example, this ratio is 30 for apricots, 40 for peaches, 160 for pears, etc.

BANANA AND PLANTAIN PROCESSING TECHNOLOGIES

Traditional Processing

Products: Uses and Dietary Significance

Most of the world's bananas are eaten either raw, in the ripe state, or as a cooked vegetable, and only a very small proportion are processed in order to obtain a storable product. This is true both at a traditional village level with both dessert and cooking bananas and when considering the international trade in dessert bananas. In general, preserved products do not contribute significantly to the diet; however, in some localised areas the products are important in periods when food are scarce. Probably the most widespread and important product is flour preparation from unripe banana and plantains by sun-drying. In Uganda, dried slices known as "mutere" are prepared for storage from green bananas, the dried slices being either used directly for cooking or after grinding into a flour. "Mutere" is used chiefly as a famine reserve and does not feature largely in the diet under normal conditions.

In Gabon, plantains are sometimes made into dried slices which can be stored and used on long journeys, and plantains are used in Cameroon to prepare dried pieces which are stored and ground as needed into flour for use in cooking a paste known as "fufu". Dried green banana slices are also used in parts of South and Central America and West Indies for preparing flour.

The other nutritionally important product is beer which is a major product in Uganda, Rwanda and Burundi where green banana utilisation is particularly high.

Preservation Methods and Processes

Drying

Both ripe and unripe bananas and plantains are normally peeled and sliced before drying, although banana figs are sometimes prepared from whole ripe. Sun drying is the most widespread technique where the climate is suitable but drying in ovens or over fires is also practiced. In west Africa, plantains are often soaked and sometimes parboiled before drying. The slices of unripe are normally spread out on bamboo frameworks; or a cemented area; or on a mat; or on a swept-bare patch of earth; or on a roof; or sometimes on stones outcrops or sheets of corrugated iron. Oven-drying of ripe bananas is practiced in Polynesia as a mean of preserving the s, which are then wrapped in leaves and bound tightly to store until needed. In East Africa a method has been reported that involves drying the peeled bananas on a frame placed over a fire for 24 hr before drying in the sun, to accelerate the process.

Product Stability and Storage Problems

There is little experimental data on the storage life of the traditionally made banana and plantain products.

Potential for Scaling up of Traditional Processes to Industrial Level

Many banana products are now produced on an industrial scale, including the traditional banana figs and flour, and the processing techniques are described below. One of the main problems encountered has been the susceptibility of banana products to flavour deterioration and discolouration and in the past many products reaching the market have been of poor quality.

A great deal of research has been directed to overcoming these problems, although however good the resultant products are they cannot compare in flavour and other characteristics with the fresh banana. Indeed, an important constraint on the large-scale development of banana processing is the lack of demand for banana products since the fresh is available throughout the year in most parts of the tropical world.

The production of beer from banana and plantains has not been scaled up to an industrial level, and while an important product in localised areas of tropical Africa, the market is rapidly declining in favour of European-type brews produced locally.

Industrial Processing

Products and Uses

The main commercial products made from bananas are canned or frozen purée, dried figs, banana powder, flour, flakes, chips (crisps), canned slices and jams. Banana products can be divided roughly into two types - those for direct consumption, such as figs, and those for use in food manufacturing industry, for example purées and powder.

Banana figs, or fingers as they are sometimes known, are usually whole, peeled carefully dried so as to retain their shape, although sometimes the is sliced or halved to facilitate drying. Banana and plantain chips (crisps) are thinly sliced pieces of fried in oil and eaten as a snack like potato chips (crisps).

The main use of canned slices is in tropical salads. Banana flakes are used as a flavouring or in breakfast cereals. Banana purée find use mainly in the production of baby foods. Banana flour is said to be highly digestible and is used in baby and invalid foods, but can also be used in the preparation of bread and beverages. Banana powder is used chiefly in the baking industry for the preparation and fillings for cakes and biscuits and is also used for invalid and baby foods.

Processing Technology

In general, to obtain a good-quality product from ripe-bananas the is harvested green and ripened artificially under controlled conditions at the processing factory. After ripening, the banana hands are washed to remove dirt and any spray residues, and peeled. Peeling is almost always done by hand using stainless steel knives, although a mechanical peeler for ripe bananas has been developed, capable of peeling 450 Kg of per hour (Banana Bulletin, 1974).

The peeling of unripe bananas and plantains is facilitated by immersing the in hot water. For example, immersion in water at 70-75 ° C for 5 min. has been suggested as an aid for peeling green bananas for flour production, while the peeling of green bananas for freezing has been facilitated by immersion in water at 93° C for 30 min.

Banana Figs

Fully ripe s with a sugar content of about 19.5% are used and are treated with sulphurous acid after peeling, then dried as soon as possible afterwards. Various drying systems have been described using temperatures between 50 and 82° C for 10 to 24 hr to give a moisture content ranging from 8 to 18% and a yield of dried figs of 12 to 17% of the fresh banana on the stem. One factory in Australia uses a solar heat collector on the roof to augment the heat used for drying bananas. Bananas can also be dried by osmotic dehydration, using a technique which involves drawing water from 1/4-in. thick banana by placing them in a sugar solution of 67 to 70 deg. Brix for 8 to 10 hr. followed by vacuum-drying at 65 to 70° C, at a vacuum of 10 mm Hg for 5 hr. The moisture content of the final products is 2.5% or less, much lower than that achieved by other methods.

Banana Puree

Banana purée is obtained by pulping peeled, ripe bananas and then preserving the pulp by one of three methods: canning aseptically, acidification followed by normal canning, or quick-freezing. The bulk of the world's purée is processed by the aseptic canning technique. Peeled, ripe s are conveyed to a pump which forces them through a plate with 1/4-in. holes, then onto a homogeniser, followed by a centrifugal de-aerator, and into a receiving tank with 29in. vacuum, where the removal of air helps prevent discolouration by oxidation. The purée is then passed through a series of scraped surface heat exchangers where it is sterilised by steam, partially cooled, and finally brought to filling temperature. The sterilised purée is then packed aseptically into steam-sterilised cans which are closed in a steam atmosphere.

Banana Slices

Several methods for canning of banana slices in syrup are used. Best-quality slices are obtained from at an early stage of ripeness. The slices are processed in a syrup of 25 deg. Brix with pH about 4.2, and in some processes calcium chloride (0.2%) or calcium lactate (0.5%) are added as firming agents. A method for producing an intermediate-moisture banana product for sale in flexible laminate pouches has been developed. Banana slices are blanched and equilibrated in a solution containing glycerol (42.5%), sucrose (14.85%), potassium sorbate (0.45%), and potassium metabisulphite (0.2%) at 90 deg. C for 3 min. to give a moisture content of 30.2%.

Banana Powder

In the manufacture of banana powder, fully ripe banana pulp is converted into a paste by passing through a chopper followed by a colloid mill. A 1 or 2 % sodium metabisulphite solution is added to improve the colour of the final product. Spray- or drum-drying may be used, the latter being favoured as all the solids are recovered.

A typical spray dryer can produce 70 kg powder per hour to give yields of 8 to 11% of the fresh, while drum-drying gives a final yield of about 13% of the fresh. In the latter method the moisture content is reduced to 8 to 12 % and then further decreased to 2 % by drying in a tunnel or cabinet dryer at 60° C.

Banana Flour

Production of flour has been carried out by peeling and slicing green, exposure to sulphur dioxide gas, then drying in a counter-current tunnel dryer for 7 to 8 hr. with an inlet temperature of 75° C and outlet temperature of 45° C, to a moisture content of 8%, and finally milling.

Banana Chips (Crisps)

Typically, unripe peeled bananas are thinly sliced, immersed in a sodium or potassium metabisulphite solution, fried in hydrogenated oil at 180 to 200° C, and dusted with salt and an antioxidant. Alternatively, slices may be dried before frying and the antioxidant and salt added with the oil. Similar processes for producing plantain chips have been developed.

Banana Beverages

In a typical process, peeled ripe is cut into pieces, blanched for 2 min. in steam, pulped and pectolytic enzyme added at a concentration of 2 g enzyme per 1 kg pulp, then held at 60 to 65° C and 2.7 to 5.5 pH for 30 min.

In a simpler method, lime is used to eliminate the pectin. Calcium oxide (0.5%) is added to the pulp and after standing for 15 min. this is neutralised giving a yield of up to 88% of a clear, attractive juice. In another process banana pulp is acidified, and steam-blanched in a 28-in Hg vacuum which ensures disintegration and enzyme inactivation. The pulp is then conveyed to a screw press, the resulting purée diluted in the ratio 1:3 with water, and the pH adjusted by further addition of citric acid to 4.2 to 4.3, which yields an attractive drink when this is centrifuged and sweetened.

Jam

A small amount of jam is made commercially by boiling equal quantities of and sugar together with water and lemon juice, lime juice or citric acid, until setting point is reached.

Product Stability and Spoilage Problems

All dried banana products are very hydroscopic and susceptible to flavour deterioration and discolouration, but this can be overcome to some extent by storing in moisture-proof containers and sulphiting the before drying to inactivate the oxidases. The dried products are also liable to attack by insects and moulds if not stored in dry conditions, although disinfestation after drying by heating for 1 hr to 80° C or by fumigation with methyl bromide ensures protection against attack. Banana powder is said to be stored for up to a year commercially and flakes have been stored in vacuum-sealed cans with no deterioration in moisture, colour or flavour for 12 months.

Banana chips tend to have a poor storage life and to become soft and rancid. However, chips treated with an antioxidant have been stored satisfactorily at room temperature in hermetically sealed containers up to 6 months with no development of rancidity.

Quality Control Methods

In general a good quality product is obtained if is harvested at the correct stage of maturity and, where appropriate, ripened under controlled conditions. For example, in the case of banana figs, the should be fully mature (sugar content of 19.5% or above) or the final. Alternatively, slices may be dried before frying and the antioxidant and salt added with the oil. Similar processes for producing plantain chips have been developed.

Banana Beverages

In a typical process, peeled ripe is cut into pieces, blanched for 2 min. in steam, pulped and pectolytic enzyme added at a concentration of 2 g enzyme per 1 kg pulp, then held at 60 to 65° C and 2.7 to 5.5 pH for 30 min.

In a simpler method, lime is used to eliminate the pectin. Calcium oxide (0.5%) is added to the pulp and after standing for 15 min. this is neutralised giving a yield of up to 88% of a clear, attractive juice. In another process banana pulp is acidified, and steam-blanched in a 28-in Hg vacuum which ensures disintegration and enzyme inactivation. The pulp is then conveyed to a screw press, the resulting purée diluted in the ratio 1:3 with water, and the pH adjusted by further addition of citric acid to 4.2 to 4.3, which yields an attractive drink when this is centrifuged and sweetened.

Jam

A small amount of jam is made commercially by boiling equal quantities of and sugar together with water and lemon juice, lime juice or citric acid, until setting point is reached.

Product Stability and Spoilage Problems

All dried banana products are very hydroscopic and susceptible to flavour deterioration and discolouration, but this can be overcome to some extent by storing in moisture-proof containers and sulphiting the before drying to inactivate the oxidases.

The dried products are also liable to attack by insects and moulds if not stored in dry conditions, although disinfestation after drying by heating for 1 hr to 80° C or by fumigation with methyl bromide ensures protection against attack. Banana powder is said to be stored for up to a year commercially and flakes have been stored in vacuum-sealed cans with no deterioration in moisture, colour or flavour for 12 months. Banana chips tend to have a poor storage life and to become soft and rancid. However, chips treated with an antioxidant have been stored satisfactorily at room temperature in hermetically sealed containers up to 6 months with no development of rancidity.

Quality Control Methods

In general a good quality product is obtained if is harvested at the correct stage of maturity and, where appropriate, ripened under controlled conditions. For example, in the case of banana figs, the should be fully mature (sugar content of 19.5% or above) or the final product is liable to be tough and lacking in flavour. However, if over-ripe is used, the figs tend to be sticky and dark in colour, so the must be fully yellow but still firm.

For banana flour, which is prepared from unripe bananas, the is harvested at three-quarters the full-ripe stage and is processed within 24 hr. prior to the onset of ripening. If less mature is used, the flour tastes slightly astringent and bitter due to the tannin content. Bananas harvested between 85 and 95 days after the emergence of the inflorescence, with a pulp-to-peel ratio of about 1.7, were considered to be most suitable for the deep-fat frying.

Other criteria suggested for assessing maturity were beta-carotene and reducing sugar content, both of which increase with increasing maturity and pH which decreases as the ripens, and these should be, respectively, about 2000 μg/100 g, less than 1.5% and 5.8 or above. Browning was found to occur if the sugar content was higher than 1.5%. The determination of crude fat in processed chips is also considered to be a necessary quality control measure. It is important to remove all impurities prior to processing of products, and this is done by washing to remove dirt and spray residues and control on the processing line so that substandard can be removed.

Preparation Methods for Fresh Bananas and Plantains

The main ways of preparing fresh bananas for consumption are boiling or steaming, roasting or baking and frying. Boiling followed by pounding into "fufu" is also widely adopted in certain areas of the tropics.

Boiling or Steaming

Plantains and bananas are often prepared simply by boiling in water, either in their peel or after peeling, and either ripe or unripe; if unripe, the is scraped thoroughly after peeling to remove all traces of fibrous material. The boiled is eaten alone or more usually accompanied by a sauce. This preparation technique is widely used in West Africa.

Roasting or Baking

Unpeeled or peeled, either ripe or unripe, is roasted simply by placing in the ashes of a fire or in an oven. This method is widely used in West Africa, East Africa and the South Pacific islands. For example, ripe plantains are placed unpeeled in an oven and when partly brown and tender, removed and peeled, then replaced in the oven and roasted evenly.

Frying

Ripe or unripe plantains or bananas are often peeled, sliced and cooked in oil, particularly in West Africa and in parts of South America and the West Indies. Similar products are also made in East Africa. Typically, ripe plantains are peeled, cut into slices or split lengthways, and fried in palm oil or with groundnut oil, the pieces being served either hot with a sauce or with fried eggs, or cold as a snack.

Pounding

Pounding is a process, used particularly in West Africa, for preparing most perishable staple food crops including plantains, cassava, yams and cocoyams to obtain a paste or dough known as "fufu" (also spelled "foofoo", "foutou", "foufou"). The plantains are peeled or boiled and peeled after boiling and pounded in a wooden mortar, the resulting paste normally being eaten with soup or a spiced sauce of meat and vegetables, but sometimes after wrapping in leaves and steaming.

Mango and guava processing technologies

Mango Processing Technologies

Mangoes are processed at two stages of maturity. Green is used to make chutney, pickles, curries and dehydrated products. The green should be freshly picked from the tree. that is bruised, damaged, or that has prematurely fallen to the ground should not be used. Ripe mangoes are processed as canned and frozen slices, purée, juices, nectar and various dried products. Mangoes are processed into many other products for home use and by cottage industry.

The mango processing presents many problems as far as industrialization and market expansion is concerned. The trees are alternate bearing and the has a short storage life; these factors make it difficult to process the crop in a continuous and regular way. The large number of varieties with their various attributes and deficiencies affects the quality and uniformity of processed products.The lack of simple, reliable methods for determining the stage of maturity of varieties for processing also affects the quality of the finished products.

Many of the processed products require peeled or peeled and sliced. The lack of mechanised equipment for the peeling of ripe mangoes is a serious bottleneck for increasing the production of these products.

A significant problem in developing mechanised equipment is the large number of varieties available and their different sizes and shapes. The cost of processed mango products is also too expensive for the general population in the areas where most mangoes are grown. There is, however, a considerable export potential to developed countries but in these countries the processed mango products must compete with established processed s of high quality and relatively low cost.

Green Mango Processing

Pickles

The optimum stage of maturity should be determined for each variety used to make pickles. There are two classifications of pickles - salt pickles and oil pickles. They are processed from whole and sliced with and without stones. Salt is used in most pickles.

The many kinds of pickles vary mainly in the proportions and kinds of spices used in their preparation. One basic recipe for the study of the preparation and storage of pickles in oil is as follows:

Mango pieces	250 g	Tumeric powder	2 to 4 g
Salt	60 g	Fenugreek seeds	2 to 4 g
Mustard powder	30 g	Bengal gram seeds	2 to 4 g
Chili powder	20 g	Gingelly oil	20 to 30 g

The ingredients are mixed together and filled into wide-mouthed bottles of 0.5 kg capacity. Three days later the contents are thoroughly mixed and refilled into the bottles. Extra oil is added to form a 1-2 cm layer over the pickles.

Chutney

The product is prepared from peeled, sliced or grated unripe or semi-ripe by cooking the shredded with salt over medium heat for 5 to 7 minutes, mixed and then sugar, spices and vinegar are added. Cook over moderate heat until the product resembles a thick purée, add remaining ingredients and simmer another 5 min. Cool and preserve in sterilised jars.

Spices usually include cumin seeds, ground cloves, cinnamon, chili powder, ginger and nutmeg. Other ingredients such as dried s, onions, garlic and nuts may be added.

Drying/dehydration. immature is peeled and sliced for sun-drying. The dried mango slices can be powdered to make a product called amchoo. The use of blanching, sulphuring and mechanical dehydration gives a product with better colour, nutrition, storability and fewer microbiological problems.

Ripe Mango Processing

Puree

Mangoes are processed into purée for re-manufacturing into products such as nectar, juice, squash, jam, jelly and dehydrated products. The purée can be preserved by chemical means, or frozen, or canned and stored in barrels. This allows a supply of raw materials during the remainder of the year when fresh mangoes are not available.

It also provides a more economical means of storage compared with the cost of storing the finished products, except for those which are dehydrated, and provides for more orderly processing during peak availability of fresh mangoes.

Mangoes can be processed into purée from whole or peeled. Because of the time and cost of peeling, this step is best avoided but with some varieties it may be necessary to avoid off-flavours which may be present in the skin. The most common way of removing the skin is hand-peeling with knives but this is time-consuming and expensive. Steam and lyepeeling have been accomplished for some varieties.

Several methods have been devised to remove the pulp from the fresh ripe mangoes without hand-peeling. A simplified method is as follows: the whole mangoes were exposed to atmospheric steam for 2 to 2 1/2 min in a loosely covered chamber, then transferred to a stainless steel tank.

The steam-softened skins allowed the to be pulped by a power stirrer fitted with a saw-toothed propeller blade mounted 12.7 to 15.2 cm below a regular propeller blade. The pulp is removed from the seeds by a continuous centrifuge designed for use in passion extraction. The pulp material is then passed through a paddle pulper fitted with a 0.084 cm screen to remove fibre and small pieces of pulp.

Mango purée can be frozen, canned or stored in barrels for later processing. In all these cases, heating is necessary to preserve the quality of the mango purée. In one process, purée is pumped through a plate heat exchanger and heated to 90°C for 1 min and cooled to 35° C before being filled into 30 lb tins with polyethylene liners and frozen at -23.50 C.

In an other process, pulp is acidified to pH 3.5, pasteurized at 90°C, and hot-filled into 6 kg high-density bulk polyethylene containers that have been previously sterilised with boiling water. The containers are then sealed and cooled in water. This makes it possible to avoid the high cost of cans.

Wooden barrels may be used to store mango pulp in the manufacture of jams and squashes. The pulp is acidified with 0.5 to 1.0% citric acid, heated to boiling, cooled, and SO2 is added at a level of 1000 to 1500 ppm in the pulp. The pulp is then filled into barrels for future use.

Slices

Mango slices can be preserved by canning or freezing, and recent studies have shown the feasibility of pasteurized-refrigerated and dehydro-canned slices. The quality of the processed product in all of these procedures will be dependent upon selection of a suitable variety along with good processing procedures. Thermal process canning of mango slices in syrup is the most widely used preservation method.

Beverages

The commercial beverages are juice, nectar and squash. Mango nectar and juice contain mango purée, sugar, water and citric acid in various proportions depending on local taste, government standards of identity, pH control, and composition of the variety used.

Mango squash in addition to the above may contain SO2 or sodium benzoate as a preservative. Other food grade additives such as ascorbic acid, food colouring, or thickeners may be used in mango beverages.

A short description of finished products found in literature is as follows:

- Mango juice: prepared by mixing equal quantities of pulp (purée) and water together and adjusting the total soluble solids (TSS) and acidity to taste (12 to 15% TSS and 0.4 to 0.5% acidity as citric acid);
- Mango nectar containing 25% purée can be prepared using the following procedure.

Brix of puree

Nectar components	15°	17°	20°
Purée	100	100	100
Sugar	45	43	40
Water	255	257	260

Commercial processing conditions may require the use of a preservative.The pH is adjusted to approximately 3.5 by adding citric acid as a 50% solution.

The time of heat processing will vary with filling temperature, can size and viscosity of the juice or nectar.

Mango squash may be prepared according to flow-sheet described below; the finished product may contain 25% juice, 45% TSS and 1.2 to 1.5% acidity and may be preserved with sulphur dioxide (350ppm) or sodium benzoate (1000 ppm) in glass bottles. Mango squash simplified flow-sheet.

Ingredients

Mango pulp	900	900
Sugar	900	1100
Citric acid	18	15
Water	900	900

Mangoes are washed, stored, peeled with stainless steel knives. The pulp is prepared by using a pulper with fine sieve (0.025-in); Sugar is mixed with water and citric acid = syrup; The pulp is added to the syrup and mixed well; The mixture is strained trough cloth; The squash is heated at 85° C and bottles are filled and closed.

For additional heat treatment bottles may need to be maintained at a product temperature of 80°C for 30 minutes if the product is to be processed without preservatives. The bottles are then left to cool in water and stored at room temperature.Two negative points must be avoided: presence of air bubbles (which is a source of quick deterioration) and separation of squash solids (giving an undesirable appearance). The means to avoid these two phenomena are described in the juices section.

A type of "squash type" beverage may also be manufactured with '/a pulp + '/a water + i/a sugar and pH adjusted to 3.7 by addition of citric acid. Using different sieve sizes affects the quality and reduces air bubbles to a certain extent but homogenisation and de-aeration of purée or squash seem to be important in order to avoid separation and air bubbles. The squash quality is evaluated on the basis of the following characteristics: pH, titrable acidity, soluble solids, ascorbic acid (by 2,6 dichlorophenol indophenol method), specific gravity.

Dried/Dehydrated

Ripe mangoes are dried in the form of pieces, powders, and flakes. Drying procedures such as sun-drying, tunnel dehydration, vacuum-drying, osmotic dehydration may be used. Packaged and stored properly, dried mango products are stable and nutritious.One described process involves as pre-treatment dipping mango slices for 18 hr (ratio 1:1) in a solution containing 40° Brix sugar, 3000 ppm SO2, 0.2% ascorbic acid and 1% citric acid; this method is described as producing the best dehydrated product. Drying is described using an electric cabinet through flow dryer operated at 60° C. The product showed no browning after 1 year of storage.

Drum-drying of mango purée is described as an efficient, economical process for producing dried mango powder and flakes. Its major drawback is that the severity of heat preprocessing can produce undesirable cooked flavours and aromas in the dried product. The drum-dried products are also extremely hydroscopic and the use of in-package desiccant is recommended during storage.

Canning

This preservation technology is described in various technological flow-sheets in this bulletin. Mango bar or " leather" is presented in various flow-sheets.

Guava Processing Technologies

Guava Purée

Guava purée is used in the manufacture of guava nectar, various juice drink blends and in the preparation of guava jam. The washed sound is first passed through a chopper or slicer to break up the and this material is fed into a pulper. The pulper will remove the seeds and fibrous pieces of tissue and force the reminder of the product through a perforated stainless steel screen. The holes in the screen should be between 0.033 and 0.045 in. The machine should be fed at a constant rate to ensure efficient operation.

The puréed material coming from the pulper is next passed through a finisher. The finisher is equipped with a screen containing holes of approximately 0.020 in. The finisher will remove the stone cells from the and provide the optimum consistency to the product.Perhaps the best way to preserve the guava purée is by freezing and the material passing through the finisher can be packaged and frozen with no further treatment. It is not necessary to heat the product to inactivate enzymes or for other purposes. The material can be frozen in a number of types of cartons and cans; however, a fibre box with a plastic bag inside is commonly used and is probably the less expensive.

It is also possible to can and heat process the guava purée and this can be accomplished by heating the purée to 195° F in an open double bottom kettle, filling into cans, closing the cans, inverting the cans for a few seconds, followed by cooling. Cans should be cooled rapidly to approximately 100° F before they are cased and stacked into warehouses.

Guava juice and Concentrate

Guava juice can be used in the manufacture of a clear guava jelly or in various drinks. A clear juice may be prepared from guava purée that is depectinised enzymatically. About 0.1% pectin-degrading enzyme is mixed into the purée at room temperature; heating of the product at approximately 120° F will greatly speed the action of the enzyme. After 1 hr. clear juice is separated from the red pulp by centrifuging or by pressing in a hydraulic juice press. A batch-type or continuous-flow centrifuge can be used on the depectinised purée with no further treatment. The clear juice after centrifuge or after press (and subsequent filtration) can be preserved by freezing or by pasteurization in hermetically sealed cans. For shipment to overseas markets it may be advantageous to concentrate either the purée or the juice.

RECENT TRENDS IN AND VEGETABLE PROCESSING

New Products

The number and variety of and vegetable products available to the consumer has increased substantially in recent years. The and vegetable industry has undoubtedly benefited from the increased recognition and emphasis on the importance of these products in a healthy diet.

Traditional processing and preservation technologies such as heating, freezing and drying together with the more recent commercial introduction of chilling continue to provide the consumer with increased choice. This has been achieved by new heating (e.g. UHT, microwave, ohmic) and freezing (e.g. cryogenic) techniques combined with new packaging materials and technologies (e.g. aseptic, modified atmosphere packaging).

The overall trend in new and vegetable products is "added value", thus providing increased convenience to the consumer by having much greater variety of ready prepared and vegetable products. These may comprise complete meals or individual components. The suitability of products and packages for microwave re-heating has been an important factor with respect to added convenience. and vegetable product trends

Heat processed products

1. Canned s and vegetables
   — Combination of vegetables in sauces and vegetable recipe dishes. Exotic s.
2. Glass packed s and vegetables
   — "Condiverde"/"antipasti" products based on vegetables in oil.
   — High quality packs.
3. Retortable plastics
   — Basic vegetables or vegetable meals
   — in jelly
4. Aseptic cartons
   — Ready made jelly
5. Rosti meals
   — Potato based products in retort pouches
6. juices
   — New combinations of juices and freshly squeezed products
7. Crisps
   — Thick and crunch skin-on crisps. Kettle or pan fried chips. Lower fat crisps.

Table. Numbers of New and Vegetable Products

	1990	1991	1992 Jan-June
Frozen vegetable products	66	95	21
Chilled vegetable products	76	81	78
Heat processed vegetables	51	60	38
Heat processed s	13	14	5
juices and drinks	73	83	46
Potato crisps	32	33	16

New product development in the and vegetable sector is most important in meeting the continued challenge of providing the consumer with choice and high quality products.

A Fresh Look at Dried

New varieties and advance in drying technologies are putting a fresh twist on dried applications. s that have been introduced to the drying process include cranberries, blueberries, cherries, apples, raspberries and strawberries - not to mention the traditional mainstays of raisins, dates, apricots, peaches, prunes and figs.Perceived as a "value-added" ingredient, dried adds flavour, colour, texture and diversity with little alteration to an existing formula. The growing interest in ethnic cuisines in U.S.A. and the change to a more healthy way of eating, has also moved dried considerably closer to the mainstream.

Found primarily in the baking industry, dried is coming into its own in various food products, including entrees, side dishes and condiments. Compotes, chutneys, rice and grain dishes, stuffings, sauces, breads, muffins, cookies, deserts, cereals and snacks are all food categories encompassing dried.

Since some dried is sugar infused (osmotic drying), food processors can decrease the amount of sugar in formula - this is especially the case in baked products. Processors are making adjustments in moisture content of the dried so that a varied range is available for different applications. An added bonus is dried s' shelf stability (a shelf life of at least 12 months). Dried is more widely available in different forms, including whole dried, cut, diced and powders.

Citric Acid and its Use in and Vegetable Processing

Citric acid may be considered as "Nature's acidulant". It is found in the tissues of almost all plants and animals, as well as many yeasts and moulds. Commercially citric acid is manufactured under controlled fermentation conditions that produce citric acid as a metabolic intermediate from naturally-occurring yeasts, moulds and nutrients. The recovery process of citric acid is through crystallization from aqueous solutions.

Citric acid is widely used in carbonated and still beverages, to impart a fresh- "tanginess". Citric acid provides uniform acidity, and its light y character blends well and enhances juices, resulting in improved palatability. The amount of citric acid used depends on the particular desired flavour (e.g., High-acid: lemonade; Medium-acid: orange, punch, cherry; Low-acid: strawberry, black cherry, grape).

Sodium citrate is often added to beverages to mellow the tart taste of high acid concentrations. It provides a cool, distinctive smooth taste and masks any bitter aftertaste of artificial sweeteners. In addition, it serves as a buffer to stabilise the pH at the desired level. The high water solubility of citric acid (181 g/100 ml) makes it an ideal additive for fountain syrups and beverages concentrates as a flavour enhancer and microbial growth inhibitor (preferably at pH < 4.6).

In processed s and vegetables, citric acid performs the following functions:

a. It reduces heat-processing requirements by lowering pH: inhibition of microbial growth is a function of pH and heat treatment. Higher heat exposure and lower pH result in greater inhibition. Thus the use of citric acid to bring pH below 4.6 can reduce the heating requirements. In canned vegetables, citric acid usage is greatest in tomatoes, onions and pimentos. For tomato packs, the National Canners Association recommends a pH of 4.1 to 4.3. In general, 0. 1% citric acid will reduce the pH of canned tomatoes by 0.2 pH units.
b. Optimise flavour: citric acid is added to canned s to provide for adequate tartness. Recommended usage level is generally less then 0.15%.
c. Supplement antioxidant potential: citric acid is used in conjunction with antioxidants such as ascorbic and erythorbic acids, to inhibit colour and flavour deterioration caused by metal-catalysed enzymatic oxidation. Recommended usage levels are generally 0.1% to 0.3% with the antioxidant at 100 to 200 ppm.
d. Inactivate undesirable enzymes: oxidative browning in most s and vegetables is catalysed by the naturally present polyphenol oxidase. The enzymatic activity is strongly dependent on pH.

Addition of citric acid to reduce pH below 3 will result in inactivation of this enzyme and prevention of browning reactions.

Cherry and Apricot Oils are Safe for Food Use

The oils obtained by cold pressing the kernels of the cherry (Prunus cerasus) and the apricot (Prunus armeniaca) have been declared acceptable for food use by the UK Ministry of Agriculture, Fisheries & Food's Advisory Committee on Novel Foods and Processes (ACNFP) by June 1993.

In its assessments of the safety in use of the cherry and apricot kernel oils, the ACNFP consider specifications that included data on fatty acid composition, the presence of natural antioxidants and the content of cyanide, mycotoxins and heavy metals.

The Committee says that it gave particular consideration to the possible presence in the oils of the cyanogenic glucoside amygdalin, from which cyanide is released by enzymic action when the kernels of cherry and apricot are crushed. Amygdalin was found to be absent from the cherry and apricot kernel oils.The ACNFP is satisfied that there are no food safety reasons why the use of cherry and apricot kernel should not be acceptable provided there is compliance with the specifications shown in Table given below.

Table. Specification of Purity for Cherry and Apricot Kernel Oils as Determined by UK ACNFP

	Cherry	Apricot
Contaminants limits:		
Heavy metals (total)	0.5 mg/kg	0.5 mg/kg
Aflatoxins (total)	4.0 g/kg	4.0 g/kg
Cyanide	0.15 mg/kg	0.15 mg/kg
Pesticide residues	0.01 mg/kg	0.01 mg/kg
Tocopherols:		
Alpha/delta/gamma (mg/kg)	356-886	569-899

The oils are obtained by the mechanical mincing and cold pressing of kernels extracted from cleaned cherry or apricot stones. After filtering, the oils are stored and are to be sold in a raw, unrefined state. The cherry and apricot kernel oils are high unsaturated and are expected to be used as speciality oils for salad dressings, baking and shallow frying applications.

The Use of Juices in Confectionery Products

During the last decade, the concept of juices has gained immensely on consumer popularity. The majority of new non-alcoholic and alcoholic drink products were a combination of syrups, juices and flavours.

The confectionery industry followed suit and new products incorporated juices as part of their confectionery formulations and processes. juice concentrates of high solids are often used instead of normal or single-fold juices.Juice concentrates are made of pure juices. The process starts with pressing s and obtaining pure juice; this is stabilised by heat treatment which inactivates enzymes and micro-organisms. The next processing step is concentration under vacuum up to 40-65° Brix or 4-7 fold. The concentrates are then blended for standardisation and stored.

These juice concentrates are often further stabilised by the addition of sodium benzoate and potassium sorbate and are usually stored away from light and are refrigerated or frozen.

Depectinised juices are also used to prevent foaming in confectionery processes and are essential for use in clear beverage products. juice concentrates which are depectinised, and have added preservatives are called stabilised, clarified, juice concentrates. juices are used in confectionery products in conjunction with natural and artificial flavours which provides intense flavour impact and are cost-effective for a confectionery product.

The traditional concern in using juice concentrates in confectionery applications has been the effect of the natural acids on the finished product, particularly the formation of invert sugar during processing.This is a logical concern since concentrates contain differing amounts and types of acids. For example: apple, cherry, strawberry and other berries contain primarily malic acid. Grapes mainly contain tartaric acid. Cranberry is high in quinic acid. Citrus s and pineapple contain differing amounts of citric acid. The concentrates, when used, are normally buffered to a pH of 5-7 with sodium hydroxide.In formulating products with juice concentrates, the solids of the concentrate are considered as mostly reducing sugars and a reduction in corn syrup is made to compensate for equivalent amount of reducing sugar being added in the concentrate.

The exact replacement can be determined by measuring the D.E. of the concentrate to be added. In formulations when small amounts of concentrate are used (less than 1%), no adjustment is made since the reducing sugar contribution of the concentrate is not significant. juice concentrates can also be used to provide a source of natural colour, in particular red colour. Grape, raspberry, cherry, strawberry and cranberry concentrates in small amounts are very effective in colouring cream centres.

The inclusion of juices in confectionery products is now left up to the imagination of the manufacturer. These products must, of course, hold up to the standards of flavour integrity, and product excellence, during the shelf-life of these products.

2

Vegetable Pest Management

INSECT PESTS OF VEGETABLES

Correctly identifying the insects and other pests that attack vegetables is the first step toward controlling these pests effectively. The pest descriptions in this booklet are designed to help the commercial vegetable grower and home gardener recognize the insects found on vegetables. Information on insect life cycles and habits is included to make pest control recommendations more understandable and usable. More specific descriptions of insects and detailed information on control procedures are available through local county Agricultural Extension Services offices.

Most of the insects considered common vegetable pests undergo a developmental process known as metamorphosis, which simply means that the insect changes form during its life. Metamorphosis may be complete or incomplete. Complete metamorphosis consists of four stages — egg, larva, pupa, and adult.

All insect life cycles include an egg stage. The insect is immobile during this stage. A larva hatches from the egg. The larval stage is the growth stage and frequently the pest stage of insect life. This is especially true of moths, butterflies, and true flies. Larvae of moths and butterflies are commonly known as caterpillars. Maggots are the larvae of true flies. Many beetles exist as grubs during the larval stage. Larvae vary in size from as short as 1/8 inch to as long as 4 inches.

The larva molts, or changes and enters the pupal stage. As during the egg stage, the insect is immobile during this pupal or changeover period. It is during this stage that the insect develops into an adult.

Adults may or may not feed in the same way and on the same plants as larvae. Beetles, for example, have chewing mouthparts and, like many grubs (beetle larvae), are capable of causing great damage to vegetables. Moths and butterflies, on the other hand, have mouthparts designed for siphoning and are unable to damage plants.Incomplete metamophorsis consists of three developmental stages — egg, nymph, and adult. The egg stage is similar whether an insect undergoes complete or incomplete metamorphosis. Immature insects called nymphs hatch from eggs. Nymphs damage plants and reduce vegetable quality by sucking plant juices. Some insects damage plants not only by extracting juice but also by injecting saliva into the plant. The best known example of this is the injury called hopperburn caused by the potato leafhopper on Irish potatoes.

The pests discussed in this booklet are divided into the following groups: Moths, Butterflies, and Their Young (Caterpillars); Beetles; Sap-Feeding Insects (True Bugs, Aphids, Leafhoppers, and Whiteflies); Flies; Other Insects; and Noninsect Pests.

MOTHS, BUTTERFLIES, AND THEIR YOUNG

Caterpillars, the larval stage of moths and butterflies, damage both the foliage and fruit of a number of vegetables. These insects chew holes in foliage and fruit and leave degrading excrement and silk on plants.

Cabbage Looper

This caterpillar feeds on the underside of leaves of cabbage and related crops and on lettuce, spinach, beets, peas, and tomatoes. It chews irregular holes through leaves. Moths lay greenish-white eggs on the upper surface of leaves. Larvae, or worms, are 1-1/2 inches long when full grown. Larvae are pale green with four thin, white lines along the back and a wide, pale line on each side of the body. These lines can be seen easily, even on small larvae. The head is narrower than the thorax (body sextion behind the head). The larvae move in a measuring or looping motion, thus the insect's name. When disturbed, loopers rear upwards. Larvae go into a resting or pupal stage in a flimsy, lace-like cocoon on the plant. The adult is a brown moth with silver markings on the wings. It has a wingspread of about 1-1/2 inches. There may be three or more generations per year, the number of worms increasing greatldy with each generation. Cabbage loopers do not overwinter in North Carolina. Moths from Georgia and Florida move into the state each year near the middle of May.

Beet Armyworm

Similar to the true armyworm, this insect is a general feeder and attacks foliage, stems, and sometimes roots of a wide variety of vegetables. Among plants attacked by this pest are beets, asparagus, corn, lettuce, cabbage, tomatoes, peppers, potatoes, onions, and peas. This pest is capable of destroying young plants. The beet armyworm overwinters as a pupa, emerging as an adult in the spring. The adult is a moth with a 1-1/4 inch wingspread. The moth's grayish brown front wings are marked with lighter lines and a spot near the center of the wing. Hind wings are white with brown edges and veins. Females lay between 500 and 600 eggs over a 4- to 10-day period. Irregular egg masses are covered with scales from the moth. Mature larvae are green with a prominent dark stripe along each side. Larvae have a dark spot on each side of the thorax, are 1-1/4 inches long, and pupate in the soil. In North Carolina, there are several generations a year.

Corn Earworm: (*Also called Tomato Fruitworm and Bollworm*)

This caterpillar feeds on a wide range of vegetables and, as a result, is known by several names. It may attack almost any vegetable but does the most serious damage to corn and tomatoes. Other vegetables attacked include peppers, eggplant, beans, okra, sweet potatoes, lettuce, and cabbage.

Fresh corn silk is preferred by this pest as a location to deposit eggs. Eggs may hatch as soon as two or three days from the time they were deposited. Newly hatched larvae feed on buds and leaves or on corn silks. When they are four or five days old, larvae travel down the silks and feed on kernels at the end of the corn ear. If fresh corn silk is unavailable, moths prefer laying their eggs on tomato foliage, especially when plants are flowering. Larvae attack tomatoes from the time the fruit forms until it ripens, boring into the sides of the fruit and near the stem. Sometimes a small, black entrance hole is the only visible sign that a large worm is inside a tomato. Injury to tomatoes is most severe in the late summer and fall.

When full-grown, larvae are about 1-1/2 inches long. They may be brown, green, or pink, with lighter colored longitudinal stripes. The head is yellow, and the underside of the body is always lighter than the back. Larvae enter the soil to pupate.

Moths are cream to grayish-brown, have green eyes, and are marked with a dark spot near each wing tip. They have a wingspread of about 1-1/2 inches. Hind wings are light colored with darker areas near the edges. Moths deposit yellow eggs one at a time on plants. One female can lay as many as 3,000 eggs. There are usually three and sometimes a partial fourth generation of earworms each year in North Carolina.

Cutworm

Many kinds of cutworms attack vegetables. Asparagus, cabbage, squash, and tomatoes are particularly vulnerable. Most cutworms cut off stems of plants at or near the soil. Cutworms feed at night when temperatures are 65 degrees F or higher. During the day these insects hide just beneath the soil close to the site of the previous night's damage. This pest is capable of damage sufficient to necessitate the replanting of corn. Cutworms usually do not eat plants beyond cutting through them, although some species climb stalks and feed on the upper parts of plants, which causes less damage. When full grown, cutworms are 1-1/2 inches long. They curl up into a tight C shape when disturbed. Most cutworms are a dull color and have practically no hair. They pupate underground. Cutworms generally overwinter in the soil in the larval or pupal stage. Adults are dull colored moths with wingspreads of 1 to 1-3/4 inches. Hind wings are usually light in color. Moths lay eggs on the stems of grasses and weeds or on bare ground. There may be more than one generation of some species each year.

Diamondback Moth

Larvae of this moth feed during the cool parts of the growing season on the underside of leaves of cabbage and related crops. Larvae feed for a time after hatching as leafminers. As larvae grow, they begin feeding on the external portion of leaves and buds. Feeding holes do not penetrate the entire leaf. Larvae are greenish-yellow with black hair and are about 1/3 inch long when full grown. They wriggle actively when disturbed and drop from the plant on silken threads. Larvae pupate inside a lacy cocoon attached to the leaf of the host plant. Moths overwinter in debris of collards, cabbage, cauliflower, and related crops. They have a wingspread of about 1/3 inch. Moths are gray to brown, marked with white. When at rest the moth has white, diamond-shaped markings along the back. The hind wings are paler and have a fringe of hair on the rear margin. Moths lay small, white eggs either singly or in groups of three or four on leaves. There may be five or six generations per year in North Carolina.

European Corn Borer

This borer attacks many species of plants, but corn, peppers, and Irish potatoes are the vegetables most severely damaged. European corn borers feed on leaves and in protected areas until about half-grown, when they bore into a plant's stem or fruit. Nearly an inch long when full grown, the caterpillar is cream colored, has a brown head, and is marked with small, round, brown spots. Caterpillars molt into pupae inside the host plant. This insect overwinters as a nearly full-grown larvae in cornstalks at ground level. Adults are pale yellow moths with irregular dark bands running across the wings. The moth's wingspread is about 1 inch. Males are smaller and somewhat darker than females. Eggs are laid in groups of up to 50 on the underside of leaves and overlap to resemble small fish scales. Most of North Carolina sees at least two generations of this insect each year, while in the eastern part of the state, there are three generations most years.

Fall Armyworm

Like other armyworms, this insect may "march" from field to field. It often is called the shatterworm because of the destruction it causes to corn foliage.

Larvae also damage corn ears as well as beans, potatoes, peppers, tomatoes, spinach, cabbage, and cucumbers. Swarms of moths move northward each spring from southern Florida and areas along the Gulf Coast. Females lay about 1,000 eggs each in masses of about 150. Eggs are usually laid on grass. Larvae are very small when they hatch but reach a length of 1-1/4 inches when full grown. Larvae vary in color from light tan to green or nearly black, are decorated with thin, yellow stripes, and have a prominent inverted white Y on the front of the head. Larvae go into the soil to pupate. Adult moths have dark gray front wings that are mottled with light and dark splotches. The wingspread of the adult moth is about 1-1/4 inches. Hind wings are grayish-white. Males have a white spot near each wing tip. It appears that this insect cannot overwinter in North Carolina. Most years there are many generations of the pest.

Imported Cabbageworm: (*Also called Cabbage Butterfly*)

This caterpillar attacks cabbage, cauliflower, kale, collards, mustard, radish, and turnips. Larvae riddle outer leaves with irregularly shaped, often large holes. Caterpillars are capable of doing enough damage to severely retard plant growth. Females lay one egg per leaf on the underside of leaves. Eggs hatch into caterpillars that are 1-1/4 inches long when full grown. Caterpillars are velvety-green and are decorated with three indistinct gold stripes, one along the back and one on each side. The pupal or chrysalid stage of the insect is brownish-grey and has many projections or spines. This insect is commonly found attached to a leaf during the pupal stage. Glue at the pupa's tail and a silken loop that encircles the middle of the pupa's body hold it in place. The adult is a white butterfly with one or two black spots on each wing. This pest overwinters as a pupa either on old plant refuse or on nearby posts or buildings. There are three or four generations per year.

Pickleworm

Pickleworm larvae bore into many cucurbit fruits. Cantaloupes, cucumbers, and squash are most seriously damaged. Caterpillars feed first on foliage. As the insect grows, it begins feeding on fruit. Pickleworms do the greatest damage to growing vine tips and blossoms early in the season. Larvae are initially white with black spots and have dark heads. There is a dark indention on each side on the back of the head. Larger larvae are up to 3/4 inch long and are green. Sawdust-like excrement produced by larvae can easily be seen extruding from damaged fruit. Winter is spent in the pupal stage in a silken cocoon on crop refuse in subtropical regions of the United States. Pickleworms usually appear in North Carolina in early July. The adult is a moth with a wingspread of about 1 inch. Wings appear fragile and are marked with pale yellow centers and yellowish-brown margins.

Potato Tuberworm

This caterpillar does significant damage feeding on exposed potatoes in the field, riddling them with narrow, silk-lined burrows. Adults are small, narrow-winged, grayish-brown moths with a wingspread of 1/2 inch. Moths tend to fly in late afternoon and evening. Females deposit single eggs on foliage of tomatoes, potatoes and related plants, or on exposed tubers of Irish potatoes. Eggs hatch into pinkish-white or green caterpillars with brown heads. When full grown, caterpillars are about 1/2 inch long. Larvae mature in about three weeks and pupate in gray, silken dirt-covered cocoons on the outside of the tuber. A brush of long dark scales protrudes at the end of the body. Moths deposit clusters of two to seven eggs on fruit, vines, buds, and leaves. There are at least four generations per year in North Carolina. Cultural control is possible by planting for early maturity and destroying plants immediately after harvest, thus preventing movement of the pest to later crops.

To prevent damage, potatoes should be kept covered with soil before harvest, and harvested potatoes should be stored in insect-proof areas. Also, vines should be cut, raked, and burned two or three days before harvesting. There may be several generations of this insect per year.

quash Vine Borer

This borer can be very destructive to squash and pumpkin plants. By boring into a vine near the base, this thick, white, wrinkled brown-headed caterpillar causes entire plants to wilt. The larva overwinters 1 or 2 inches in the soil within a dark, dirt-encased cocoon and molts into a pupa in the spring. Within two or three weeks after molting, the 3/4-inch-long pupa has worked its way to the soil surface and emerged as an adult. The adult squash vine borer is a moth with a wingspread of about 1-1/2 inches. The front wings are covered with metallic, greenish black scales.

Hind wings are transparent. The abdomen is ringed with red and black. Squash vine borer moths look and act like paper wasps. Females lay red eggs near the base of plant stems. Eggs produce smooth, white caterpillars that bore into the center of the vine a few days after hatching. Small piles of green excrement and wilting of whole vines are signs a plant is infested with squash vine borers. There are two generations of this pest each year. Damage is usually noticed in early June and again in early August.

Tomato Hornworm and Tobacco Hornworm

Tomato and tobacco hornworms feed on tomatoes, peppers, potatoes, and related plants. These green worms are 3 to 4 inches long and can defoliate young plants. Mature tomato hornworms can be distinguished by eight L-shaped, white markings along each side of the body and a bluish-black horn near the rear end of the insect. Tobacco hornworms are decorated with diagonal white bars and have a red horn. The white cocoons of a parasite are usually attached to the worm. Pupae of both species are mahogany brown and are found in the soil. In both species, adults are large, gray moths with mottled front wings. Hind wings are lighter. Two thin, zigzag lines run across the midsection of each hind wing of tomato hornworm moths. There is a dark area between these somewhat obscure lines on the wings of tobacco hornworm moths. On both species, five or six yellowish-orange spots can be found along each side of the body. Moths of both species have a wingspread of 4 to 5 inches and fly mainly around dusk. Females lay eggs individually on plants. Eggs are spherical and greenish-white. This insect overwinters as a pupa. In North Carolina, there are two generations a year.

True Armyworm

This caterpillar is sometimes present in very high numbers. Armyworm epidemics can result in the destruction of large areas of vegetation. This pest attacks grasses, corn, and some legumes. Damage usually starts at the edge of a field. Larvae feed at night and may move in hordes. Armyworms are capable of consuming corn plants up to 8 inches tall. This insect stops feeding and overwinters as a partially grown larva. It resumes feeding in the spring. Armyworms are dark greenish-brown worms up to 2 inches long with white stripes down the sides and back. The insect pupates just below the surface of the soil.

Yellowstriped Armyworm

This caterpillar is a daytime feeder. It is capable of doing significant damage to young tomato plantings, pepper fruit, and, occasionally in August, to sweet potatoes. Its presence is sporadic, however. Overwintering as pupae in the soil, yellowstriped armyworms emerge as moths in the spring. Adults are brown, black, and white moths with a 1-1/2 inch wingspread. The moth's forewings are marked with a diagonal band that runs from the front to the rear of the wing and a smoky white stripe along the outside edge of the wing. Hind wings are white with a dark edge. Eggs, covered with scales from the female's body, are laid in clusters on foliage, trees, and buildings. Larvae are dark green with a yellowish-orange stripe along each side and black spots on their abdominal segments. In North Carolina there are several generations a year.

BEETLES

Adult beetles are usually hard-bodied insects with thick forewings. The young are grubs, borers, or wireworms. Often adults feed on different host plants than do larvae, although both stages may be destructive to vegetables.

Asparagus Beetle

There are two beetle species that feed on asparagus — the asparagus beetle and the spotted asparagus beetle. The spotted asparagus beetle, which is reddish-orange and black, is more destructive in northern asparagus production areas. Asparagus beetles are more troublesome in North Carolina. The asparagus beetle is metallic blue with six yellow areas on the wing covers. Beetles are about 1/4 inch long and have a narrow, red head and thorax. This pest overwinters in protected areas and first appears each year in early April along the edges of fields.

Adults gnaw tender asparagus buds and lay eggs on developing spears. Eggs are very dark, bulletlike, and laid in rows. Processing plants sometimes reject otherwise edible spears because eggs have been laid on them. Eggs hatch within a week of the time they were laid. Sluggish, humpbacked, slate-colored larvae with black heads feed and grow on asparagus ferns for about 10 days. The pupal stage is passed on the soil. A week or so after pupation, adults emerge and the cycle is repeated. There may be as many as five generations in a year. Asparagus is the only vegetable this insect attacks.

Bean Leaf Beetle

Adult bean leaf beetles range in color from red to yellow. The insect's markings may also vary, but bean leaf beetles generally have a dark triangular mark on the front of the wing covers, six black spots in the center of the back and a black edge around the outside of the wing covers. This insect is about 1/4 inch long and is long-lived. Adults will drop from foliage when disturbed. Bean leaf beetles feed on foliage, making round, regular holes. This pest also attacks the stems of young plants at or slightly below ground level. Females lay eggs in the soil near the base of plants. From the eggs hatch slender, white larvae that feed on the roots of bean plants. Bean leaf beetles spend the winter as adults in trash near bean fields. There is one generation, possibly two, per year. Beans, peas, cowpeas, and soybeans are common host plants; however, this pest will also feed on corn, some clovers, and many weeds.

Blister Beetle

Several species of blister beetle feed on vegetables. The black, striped, and margined blister beetles are most commonly seen in North Carolina. Blister beetles are slender, rather soft insects with a narrow thorax. The tip of the abdomen is exposed beyond the wing covers. Adults are black or gray and sometimes have narrow, yellow or gray stripes or edges on the wings. Adult beetles are ravenous feeders and are capable of destroying plants. The winter is passed as larvae in the soil. Larvae are about 1/3 inch long and yellow. Blister beetle larvae are beneficial in that they feed on grasshopper eggs. The pupal stage also is spent in the soil. Adults emerge from the soil in midsummer and feed in groups for about five weeks. The beetle's body contains an oil called cantharidin that causes blisters if the insect is crushed against tender skin.

Yellow eggs are laid in holes in the soil in clusters of 100 or more. It appears there is one generation per year. Blister beetles will attack most types of vegetables but seem to prefer potatoes and tomatoes.

Colorado Potato Beetle

This insect is perhaps best known as the common yellow- and black-striped potato bug. This pest is, however, a beetle and not a true bug. Both adults and larvae feed on the foliage of potatoes. While potatoes are preferred, this pest also attacks tomatoes, eggplant, peppers, and horse nettle. Adults are about 3/8 inch long and 1/4 inch wide and very convex in shape. Black and yellow stripes run lengthwise along the insect's back. Females lay clusters of 20 or more yellowish-organe eggs on the underside of leaves. A single female beetle can lay over 500 eggs over the course of her four-week life. From the eggs come red, hump-backed, soft-bodied larvae decorated with two rows of black spots along each side of the body. Larvae begin chewing on leaves immediately after hatching. Larvae also feed on the fruit of tomatoes and eggplant. In two or three weeks larvae grow to a length of 1/2 inch, then enter the soil to pupate. Seven days later adult beetles emerge from the soil, feed for a few days, mate, and lay more eggs. This insect overwinters as an adult in the soil. There are at least two generations per year.

Cowpea Curculio

This insect most commonly attacks blackeyed and crowder peas. Adults are black, humpbacked snout beetles about 1/4 inch long. The adult female chews a hole in a pea pod and inserts an egg into the developing seed. This small puncture is difficult to see, and damage may go unnoticed until the crop is harvested. Eggs hatch into small, white, legless grubs with yellowish-brown heads. The grubs destroy the peas in a pod. There is at least one generation per year. Hot, dry weather appears to retard development of this pest. Adults overwinter in refuse around fields. Damage may be avoided by alternating from year to year the site where peas are planted.

Flea Beetle

The many different species of flea beetle attack a variety of vegetables. Species that attack potatoes, tomatoes, and eggplant are the most troublesome, but flea beetles also damage corn and sweet potatoes. Flea beetles are so named because of their enlarged hind legs and jumping ability. Adults are 1/16 to 1/18 inch long and usually black with brown legs and antennae. Eggs are laid in the soil at the base of plants. From the eggs hatch cylindrical, brown-headed, white larvae that are about 1/5 inch long when full grown. Larvae commonly feed on roots, often riddling them with tunnels or eating off rootlets. The pupal stage is spent in the soil. These insects overwinter as adults in trash around field borders and in ditch banks. There are three or more generations per year. Round holes chewed in leaves are the most obvious damage caused by flea beetles. On small plants, this "buckshot" damage can result in plant death. On larger plants these feeding sites provide a good entrance for several blights and wilts.

Japanese Beetle

These beetles are metallic green or greenish-bronze, 1/2 inch long with red wing covers, and are decorated with white spots near the tip of the abdomen. They commonly are found in June and July on fruit, deciduous trees, and vegetables. Japanese beetles also are frequent and persistent pests of sweet corn and beans. The insect spends the winter in the soil as a white grub about 3/4 inch long. Beetles emerge from the soil in midsummer and on warm, sunny days swarm to plants, where they eat the foliage, flowers, and fruit. Females beetles then return to grassy areas to lay their white, spherical eggs in the soil. Grubs feed on decaying vegetation and grass roots. Pupation occurs in the soil. There is only one generation each year.

Mexican Bean Beetle

The brown Mexican bean beetle is decorated with 16 black spots arranged in three rows across its back. Beetles are 1/3 inch long, convex, and resemble lady beetles. Adults overwinter on the ground among leaves and trash near fields where beans were grown. When warm weather arrives in the spring, they leave their overwintering sites and feed on beans for a week or two. Females lay lemon-colored eggs in groups of 25 or more on the underside of leaves. Eggs hatch about 10 days after being laid and yellow larvae with six rows of black-tipped, branched spines on their backs appear. Larvae, which are 1/3 inch long when full grown, feed for two to five weeks, then enter the pupal stage. The yellow pupa is attached to a leaf. Traces of the last larval skin can usually be found where the pupa is attached to the leaf. There are three or four generations per year. Both larvae and adults feed on leaves, usually on the underside. Damaged leaves have a skeletonized or lace-like appearance. Sometimes stems and pods are damaged. The insect attacks all kinds of garden beans, cowpeas, and soybeans.

Spotted Cucumber Beetles: (*Also called the Southern Corn Rootworm*)

In its adult form, this insect feeds on over 200 different plants. It may be found on beans, cucurbits, corn, potatoes, asparaugs, cabbage, peas, beets, tomatoes, and turnips. It feeds on foliage and new shoots, leaving small holes in foliage. It also gnaws holes in fruit. This insect overwinters as a 1/4 inch long, yellowish-green beetle decorated with 12 black spots. The beetle's head and antennae are black. Females deposit eggs at the base of grass and corn plants. From the eggs hatch yellowish-white, brown-headed larvae that are up to 3/4 inch long when full grown. These larvae, called corn rootworms, tunnel in the roots of plants. This tunneling action can result in severe stunting of corn plants and sometimes kills plants. There appears to be two generations per year in North Carolina.

Striped Cucumber Beetle

These 1/5-inch long beetles are yellow and are decorated with three wide black stripes. Striped cucumber beetles damage plants by feeding on leaves, stems, and fruit. They also carry the organisms that cause bacterial wilt of cucurbits and cucumber mosaic. Adults lay yellow eggs at the base of plants. Larvae, which are white and 1/3 inch long, feed on the underground parts of plants or on fruit that touches the soil. There are at least two generations per year. In addition to cucurbits, striped cucumber beetles feed on beans, peas, and corn. The most severe damage is caused to small plants.

Vegetable Weevil

Larvae and adults of this insect feed on foliage and exposed roots of potatoes, turnips, turnip greens, kale, and mustard greens. Late fall and early spring crops are most severely damaged. Adults are gray snout beetles 1/3 to 1/2 inch long. Two white markings on the rear half of the wing covers form a V shape. Eggs laid in the crowns of plants procude green, slug-like larvae. These grubs are about 1/2 inch long when full grown. This pest goes into a resting stage from June to September.

Wireworms

These beetles do the most severe damage to root crops such as Irish and sweet potatoes, although they injure other vegetables as well. There are many different species of wireworms. Adults are hard-shelled, dull colored beetles, ranging from 1/2 to 1-1/2 inches in length. Beetles are tapered toward both ends and appear streamlined. Adults are also known as click beetles because they snap and flip over when turned on their backs. They have a loose, flexible joint just ahead of the wings. Females lay eggs in the soil around the roots of grass plants. The eggs produce yellow to reddish-brown, tough-bodied larvae that are commonly called wireworms. Larvae range in size from 1/2 to 2 inches when full grown, and most are cylindrical and slender. Some wireworms are almost flat. Larvae feed on underground parts of plants. The pupal stage is spent in the soil. It takes over a year to complete the life cycle of a wireworm. Wireworm damage is most severe on land not previously planted in row crops.

White Grubs

There are several species of these pests. Adults are hard-shelled, light brown to almost black beetles from 3/4 to 1 inch long. These insects are sometimes called May beetles and June beetles. They commonly feed at night on the foliage of trees and hide in the grass and lay eggs during the day. The spherical, pearly white eggs produce white C-shaped grubs with brown heads. Grubs feed on grass roots and other crops.

The pupal stage is spent in the soil. Among the common white grubs are the larvae of the green June beetle. Larvae feed mainly on organic matter in the soil. Grubs uproot young plants as they move about in and on the soil. Vegetables should not be planted in soil that was in sod the previous year until the soil has been treated to control white grubs as some grubs live two to three years.

Whitefringed Beetle

Among the many vegetables attacked by whitefringed beetles are Irish potatoes, sweet potatoes, beans, and cabbage. Adults, which emerge in June and July, are dark gray beetles about 1/2 inch long, with a white margin or fringe on their sides. White, oval eggs are laid in a gelatinous mass on objects on or near the ground. Grubs are shite, legless, and 1/2 inch long when full grown. Grubs feed on the underground parts of many kinds of plants. The pupal stage is spent in the soil, while the insect overwinters in the grub stage. There is one generation per year.

SAP-FEEDING INSECTS

Some insects damage vegetables by sucking plant juices from leaves, stems, or roots. The true bugs, aphids, whiteflies, and leafhoppers, all have beak-like mouthparts that are used to suck sap from plants. Some of these insects also produce toxic saliva, which they inject into plants. The best known example of plant injury caused by saliva injection is a condition called hopperburn caused by the potato leafhopper on Irish potatoes. Many of these insects also transmit plant diseases. The green peach aphid, for example, is known to transmit potato leafroll virus, while the melon aphid infects squash and watermelon mosaic virus.

Harlequin Bug

These bugs feed on cabbage and related crops and can cause severe damage. Harlequin bugs also feed on beans, asparagus, okra, and tomatoes. The insect damages a plant by extracting juices from it. Adults and nymphs have similar habits. Adults are flat, shield-shaped bugs about 3/8 inch long. They are red or organe with black markings that give the appearance of a mask. Eggs resemble kegs and are laid on end in groups. Each egg has a black band at the top and bottom. Nymphs, which grow through five developmental stages, resemble adults but are smaller and wingless. There are two or more generations of this insect per year. Adults overwinter in any protected area.

Eggplant Lace Bug

Adult eggplant lace bugs are small, flat insects about 1/8 inch long. They are somewhat rectangularly shaped when viewed from above. The wings are net or lace-like. The head is covered with a hood-like structure. Females lay eggs on the underside of leaves. Nymphs are darker than the adults, wingless, and usually covered with spines. Both nymphs and adults inject saliva and suck juices from eggplant leaves. The insect also excretes a brown, varnish-like material.

Squash Bug

This insect attacks only cucurbits. It prefers squash and pumpkins. Adults and nymphs injure plants by sucking sap from them. Damage usually can be detected by the occasional drooping of a leaf. Adults are about 5/8 inch long, brownish-black or gray, and flat across the back. This insect overwinters as an adult and mates in the spring. Reddish-brown eggs are laid on leaves in rows that come together at an angle. Each cluster contains 15 or more eggs, which produce tiny, light gray, wingless nymphs. There is one generation each year. Destruction of crop residue as soon as the crop is harvested helps to control these insects.

Tarnished Plant Bug

Among the many different vegetables this insect attacks are asparagus, Irish potatoes, chard, celery, and beans. A small spot surrounded by a yellow area is evidence a tarnished plant bug has been feeding on a bean seed. Blossoms may drop from a plant as a result of tarnished plant bugs feeding on them. This insect also feeds on the terminals of potatoes, causing leaves near the top of the plant to wilt. Tarnished plant bugs also transmit disease. Adults are small, flat bugs about 1/4 inch long.

These somewhat oval insects are mottled brown with irregular splotches of white, yellow, reddish-brown, and black. Eggs are usually inserted into plant tissues. The nymphs resemble adults but are smaller and wingless. There appear to be four or five generations per year in North Carolina. This insect overwinters as an adult. Two other true bugs — stink bugs and leaffotted bugs — cause similar injury, especially on beans and southern peas. Damage inflicted by these two bugs on southern peas and butter beans occasionally is serious.

Aphids

Aphids, or plant lice, feed on most vegetables. These insects cause the greatest damage when they suck juices from and inject saliva into plants. Many aphid species transmit viruses that cause diseases. Adults are soft-bodied, pear-shaped, and may or may not have wings. Two cornicles, or "exhaust pipes," extend from the hind end of most aphids. Aphids may be green, pink, yellow, or black. Nymphs resembe adults but are smaller and always wingless. These pests usually feed in clusters.

Potato Leafhopper

This insect feeds on many different host plants including beans and Irish potatoes. Saliva injected by this pest into Irish potato plants causes a deadening and upcurling of the tissues known as hopperburn. Sap feeding on beans causes leaves to be stunted, crinkles, and curled downward. Adults are pale green, wedge-shaped, and about 1/8 inch long. Eggs inserted into plant tissues produce green nymphs that resemble adults but are smaller and wingless. Adults and nymphs suck plant juices. Potato leafhoppers are a problem primarily in western North Carolina. At least three generations of this pest occur each year. The potato leafhopper is one of several leafhopper species found on vegetables.

Whitefly

Whiteflies feed on a wide variety of vegetables. This insect can be found in the field and the greenhouse. Host plants include beans, tomatoes, eggplant, and cucumber. Adults are very small sucking insects with two pairs of broadly rounded wings. The wings and body are covered with a snow white, waxy powder. The insects look like tiny moths and may fly up in a cloud when disturbed. Eggs, which are almost microscopic, are laid on the underside of leaves. Young nymphs are small, flat, and hard to see. All stages are found on the underside of leaves. Leaves may become discolored and fall from a plant on which whiteflies are feeding. Plants on which this insect feeds may also be stunted. Whiteflies secrete honeydew on plants, which allows unsightly black sooty mold to grow.

FLIES

True flies are a diverse group. Many species attack humans and animals and can spread diseases. Some flies are beneficial. They feed on decaying matter and prey on or parasitize other insects. Others are plant pests, their maggots or larvae feeding on various vegetables.

Cabbage Maggot

This insect prefers cabbage, cauliflower, broccoli, and related crops. It also attacks beets, radishes, turnips, celery, and other vegetables. The cabbage maggot does most of its damage in the mountain regions of North Carolina. The adult is a gray fly with two wings. It is smaller than a house fly but has proportionately longer legs. Eggs are white and are laid on the soil close to young plants. The eggs hatch into white, wedge-shaped maggots that are about 1/4 inch long when full grown. Maggots feed on roots, either on the surface or by tunneling through them. Fleshy parts of radishes or turnips will show brown streaks where tunneling has occurred. Cabbage maggots overwinter in the pupal stage, which is spent in the soil. There appear to be three generations per year of this pest.

Pepper Maggot

This pest prefers peppers but also attacks eggplant. It is believed that its natural food is horsenettle. Adults are yellow striped flies about 1/4 inch long with dark bars on the wings. Females deposit eggs under the skin of peppers. Maggots are white, legless, and about 1/2 inch long when full grown. Maggots eat the core of the pepper, causing it to decay or drop from the plant. The pupal stage, in which the insect overwinters, is spent buried 2 to 4 inches in the soil. Adult flies appear in late June. There is one generation each year.

Seedcorn Maggot

This insect attacks the sprouting seeds of many different vegetables. Damage may be done to seed pieces of Irish potatoes; seedlings of corn, beans, peas, and cucumbers; or stems of other vegetables. The most serious infestations occur in the spring during cool, wet weather and in soils high in organic matter.

Adults are grayish-brown flies about 1/5 inch long. Eggs, from which hatch cream-colored, wedge-shaped maggots, are deposited in the soil close to plants. Maggots are 1/4 inch long when full grown, legless, and have chewing mouthparts at the narrow end of the body. It is as a maggot that this insect is destructive. This pest overwinters in the soil in the pupal stage. The pupa is enclosed in a brown, capsule-like case. There are three generations of this pest each year. Destruction of crop refuse and shallow planting in warm soils will enhance plant establishment.

Vegetable Leafminer

This insect's name derives from the maggot's habit of tunneling or mining in leaves — feeding between the upper and lower leaf surfaces. Damage appears as winding white trails or broad white spots. Leafminers attack many vegetables but do the most damage to cucumbers, squash, tomatoes, and leafy vegetables such as cabbage, kale, and collards.

The leaves of a plant that is attacked by vegetable leafminers are weakened, while the mines serve as sites for disease and decay. Adults are gray to shiny black flies less than 1/8 inch long. Small, white eggs deposited in leaves produce white or yellow maggots that tunnel through the leaves. The tunnel is narrow and winding at first. It becomes larger as the maggot grows, finally forming a blotch. The pupal stage is spent in the soil, and it is in this stage that this pest overwinters. There are several generations each year in North Carolina.

OTHER INSECT PESTS

Moths and butterflies, beetles, sap-feeding insects, and flies are not the only insects capable of damaging vegetables. Grasshoppers, mole crickets, and thrips also attack vegetables.

Grasshoppers

Many species of grasshopper damage vegetables. Usually these pests migrate from dried up vegetation to more succulent vegetables. Adults and nymphs cause damage by feeding on foliage. Adults have front wings that are longer than the body and are held roof-like over the insect. Hind wings are softer and are folded fanlike beneath the forewings when the insect is at rest. The hind legs are long and adapted for jumping. Size varies according to species. Nymphs resemble adults in habits and appearance but are smaller and wingless. Most species overwinter as eggs, which are laid in packets in the soil.

Mole Cricket

This insect burrows in the soil, causing major damage by uprooting young plants. Mole crickets are capable of severely damaging the roots and tubers of Irish and sweet potatoes. Damage to foliage may also occur. Adults are velvety brown and about 1-1/2 inches long. This insect has short forewings and long, membranous hind wings. The front legs, rake-like and developed for digging, somewhat resemble the forepaws of a mole. Hind legs are enlarged. Eggs are laid in an underground chamber. Nymphs are small, wingless versions of adults. Nymphs and adults overwinter. There is one generation each year in North Carolina.

Thrips

Several species of thrips damage vegetables. The most severe damage usually occurs in hot, dry years. Damage appears as white flecks or streaks on leaves. Adults are small s

lender insects about 1/20 inch long. They vary in color from light to dark brown. Wings are narrow and fringed, giving them a feathery appearance. The males of some species are wingless. Eggs are thrust into plant tissues. Nymphs are active, light-colored, wingless insects. The mouthparts of adults and nymphs are similar and unusual; they are designed to rasp plant tissues then suck the juices. The last nymphal stage is spent in the soil. Adults and nymphs pass the winter on plant tissue. There may be as many as five generations each year in North Carolina.

NONINSECT VEGETABLE PESTS

Spider mites and slugs, although not insects, are capable of inflicting severe damage on vegetables.

Spider Mites or Red Spiders

Mites are closely related to insects. Many species of this pest attack vegetables. Beans and tomatoes may be severely damaged, while crops such as eggplant, cucumber, and celery are also susceptible. These pests appear to cause more severe damage during hot, dry years. Adults are usually red, yellow, or green, about 1/ 50 inch long, and have four pairs of legs. The body is oval. Adults lay eggs on plants. The eggs hatch into young mites that resemble adults but are smaller. Adults and nymphs suck plant juices. Plants attacked by mites appear sickly and white-speckled. Affected leaves curl, turn brown, and drop from the plant prematurely. When an infestation is severe, fine silk webs may entangle an entire plant, and moving spider mites can easily be seen in the webbing. Many generations may occur each year.

Slugs

Slugs are similar to snails but lack the snail's hard, protective shell. Six species of slugs attack vegetables and ornamental plants in North Carolina. Slugs feed on mold, decaying organic matter, and live plant tissue including foliage, flowers, or fruit. Slugs may chew jagged edges along leaves, shred leaves, or eat completely through leaves or stems. Damage is most severe in damp areas or during extended rainy periods. Slugs vary in size from less than 1/2 inch to more than 4 inches long and range in color from ash-gray or yellow-gray to black. Some are mottled. Young slugs resemble adults but are smaller and lighter in color. Eggs, gelatinous and water appearing, are laid in clusters of about 25 on the soil in concealed, moist areas. Slugs are active at night, leaving silvery slime trails that can be seen the next morning.

MANAGING INSECT PESTS

Several mechanical and cultural practices may be used to help reduce insect pests in home vegetable gardens. Some are more practical than others, and success will depend in part on your willingness to work at them. An integrated approach to pest management (combining strategies) is often best. One of the most important strategies in dealing with insects is to learn about their habitat, behavior, life cycle, what they feed on, and whether they are actually pests. This information will help you decide what to do. Most insects found in the garden are not pests, and some are even beneficial. The following practices help manage insect pests in vegetable gardens:

- Maintain vigorous, healthy plants. Evidence suggests that plants growing under stressful conditions are more likely to be attacked and suffer serious damage. Fertilization, liming, too little or too much water, and planting too close together can all adversely affect plants. Check the fertility and pH of soil regularly and make adjustments as needed. Thin plants to the recommended spacing.

- Planting the same crop in the same place year after year may cause pest buildup. Rotate crops, especially where soil insects such as grubs, wireworms, and maggots are a problem. Do not plant crops susceptible to grubs or wireworms where grass grew the previous year.
- Choose varieties recommended for your area and, where available, varieties resistant to pests known to occur in your area. For example, butternut squash is resistant to the squash vine borer.
- Sanitation in and around the garden is very important. Many vegetable pests overwinter on weeds or plant debris in or near the garden. Remove weeds and organic mulches, which can provide ideal homes for insects, slugs, and snails.
- Avoid bringing insect-infested plants into the garden. Carefully check transplants for the presence of insects before purchasing and planting.
- Consider altering time of planting: could the pest be avoided by planting earlier or later?
- Handpick pests off the plants and destroy them. Insects may be killed by placing them in a bucket of soapy water.
- Physical barriers placed around plants can control some insects. Barriers include:
  - Cardboard collars (or roofing paper), 4 in. high, placed around young transplants to prevent cutworms from cutting the stems. Squares of tarpaper or carpeting placed securely around the stems of young crops of the cabbage family can prevent the cabbage maggot fly from depositing eggs at the base of the plant.
  - Row covers placed over the plants until the pest is gone or the plants are large enough to need the covers removed. All covers should be removed about four to six weeks into the season because temperatures during midsummer get too hot. Remember that some plants such as cucumber, eggplant, melons, and squash need to be pollinated by insects to yield a crop. Peppers may also benefit from insect pollinators. Commercial covers made of polypropylene, polyester, or polyvinyl alcohol are available, but cheesecloth or screening can also be used. All of these covers let light and water in and allow continued plant growth. Even ventilated plastic row covers help keep out many pests.
- Mulching materials such as aluminum foil may repel aphids, thrips, and other insects. Although expensive, mulching may be practical on a small scale.
- Traps, such as yellow sticky boards, can be used to help monitor insect populations, but they are seldom sufficient to give control. They do, however, help keep whitefly populations low as long as the sticky material is replaced periodically when insects cover the boards.

- Take advantage of natural enemies, predators, and parasites. Learn to recognize those that are almost always present and conserve them. Small wasps parasitize aphids, leaving bronze or gray and bloated aphid "mummies." Immature lady beetles and lacewings, which resemble small alligators, are also frequently present. Others include spiders, predatory mites, predatory bugs, predatory flies, and ground beetles.
  - Augmentation: The introduction of predators, parasites, or diseases is becoming more practical as we learn more about managing the pest system. Remember when introducing or maintaining predators or parasites, beneficial insects will move elsewhere if there are insufficient hosts to feed on.
- Pesticides may also be used as part of the pest management program. Be sure to use only the amount you need and to treat only the crops that need treating. Spot treatments are effective and may be practical for home gardens.
  - *Note:* Even if a pesticide is botanical in origin, it may be toxic. Some botanical insecticides are more toxic than some of the commonly available synthetic chemicals.
  - Biorational pesticides such as *Bacillus thuringiensis* (Bt), a toxin produced by bacteria that kills caterpillars, are an alternative to some chemical pesticides.
  - Insecticidal soaps and horticultural oils are also an alternative to some traditional chemical pesticides and may be useful for certain pests, especially aphids, in the home garden.
  - Diatomaceous earth, a desiccant, is sometimes used to control insects, slugs, and snails. Once it gets wet and compacted, however, it loses its effectiveness.

Before using any pesticide, check the label. Both the crop you want to treat and the pest you are treating for must be listed on the label. If not, do not use the pesticide. No matter which methods you choose, keep a record of what you did and whether it was successful. Such a record should be a great help in the future when you are faced with similar pest management decisions.

MINIMIZING VEGETABLE DISEASES

To grow a healthy vegetable garden— one with few or no diseases—some general practices can be followed. The following 10 steps will maintain healthy plants and reduce the need for fungicides. You may be able to devise others that are especially suited to your garden.

- Important way to manage plant diseases. The letter abbreviations used to describe the resistance of a variety (e.g., VF = Verticillium and Fusarium wilt resistant, PM = powdery mildew resistant or tolerant) are listed in seed catalogs or can be explained by your county Cooperative Extension agent. Resistant varieties resist infection by a particular disease agent and show little or no disease. Tolerant varieties may show symptoms but still yield the same as resistant varieties or susceptible ones protected with pesticides. When available, choose varieties that are resistant or tolerant to a disease that has been a problem.
- *Purchase treated seed*. Seed may come pretreated with a dusting of a fungicide, or you may dust the seed with a fungicide. This coating will help prevent the seed from rotting in the soil before germination and can help protect the emerging seedling from damping-off. If seed rot or damping-off has been a problem in your garden, treating the seed with a fungicide will help.
- *Purchase disease-free seed, transplants, and propagating material*. Begin with healthy plant material to help plants quickly become established in the garden. Plant materials that are unhealthy to begin with never yield as much as healthy ones or may die while still young. Reputable seed companies sell only disease-free plant materials. Some seeds are hot-water treated to remove infectious agents. Some are tested to reduce the risk of seedborne viruses. When shopping for transplants or other propagating material, take time to examine the plant stock thoroughly to make sure it is healthy and vigorous. If you save your own seed, harvest it from healthy plants and dry it thoroughly. Store such seed in properly labeled, airtight containers in a cool, dry place.
- *Select a sunny, well-drained location*. A sunny area with welldrained soil is an ideal site for vigorous growth of plants. Shaded, poorly drained areas result in weak and spindly plants that are easy targets for disease organisms. Even if such plants remain alive and free of infectious disease, they will not yield as much as strong and burly plants.
- *Improve the soil environment*. When there is no other choice for a garden site but a heavy, wet soil, plant in raised beds or ridged rows so the soil around the plants' roots will be drier. Heavy, wet soils discourage healthy root growth and encourage root rots. When a garden is established on sloped terrain, plant in terraced beds to reduce soil erosion over delicate, young plants and newly sown seed.

Soils that are dry and sandy may be mulched with a variety of materials (such as straw, grass clippings, or black plastic) to help retain moisture. A soil environment that is favorable to healthy root development supports the growth of healthy plants.

- *Water and feed plants.* Plants require 1 inch of rainfall per week for best growth. If rainfall is inadequate, water the garden. Water plants in the morning so they will dry off quickly above ground, reducing the chances of disease spread. Avoid using overhead sprinkler irrigation because it can promote the development and spread of leaf, flower, and fruit infections. Trickle irrigation is best because it puts water directly in the root zone, does not wet the plants above ground, and does not encourage soil splashing. Plants that are fertilized properly at planting time and sidedressed will grow better and be healthier. Always use a complete fertilizer or incorporate a well-rotted manure or rich compost into the soil. Avoid overfertilization because this injures plant roots.
- *Space plants to allow air circulation.* High humidity and moisture favor the development of diseases on plants. Allowing enough room for plants to grow and space for air to circulate around mature plants reduces humidity and promotes rapid drying of plant surfaces. This helps reduce incidence of disease.
- *Practice cleanliness in the garden.* Always remove and destroy or discard (in the trash) plant materials that show signs of disease. Work in the garden when plants are dry because moisture on plants aids the spread of infectious diseases. Composting, unless the pile becomes very hot, does not effectively eliminate diseases from plant refuse under New York State climatic conditions. For this reason, it is unwise to compost any diseased plant material. At the end of the growing season, clean up all crop debris because disease agents overwinter in debris and may infect new plants the following season.
- *Plant a fall cover crop and plow it under the following spring.* After cleaning up the garden, sow a grass, such as perennial rye, that will begin to grow that fall. This cover crop will protect the topsoil from erosion during the winter. The following spring plow under the ryegrass to enrich the soil with fresh organic matter or "green manure." This practice also helps reduce the populations of certain soilborne disease agents. Noninfectious agents flourish on green manure in the soil and tend to inhibit the infectious ones.
- *Rotate crops.* Successive planting of one crop family in the same area for many seasons promotes the buildup of disease agents in the soil. Thus the disease becomes more severe over time. Rotate plants to different areas of the garden to help reduce the losses caused by

soilborne disease agents. Avoid successive planting within crop families or crop types such as crucifers (cabbage, broccoli, turnip, radish), cucurbits (melon, cucumber, squash), solanaceous plants (tomato, eggplant, potato, pepper), grasses (sweet corn, cover crops such as rye), legumes (bean, pea), and root crops (carrot, beet, onion).

TROUBLESHOOTING IN THE HOME GARDEN

The following are common problems many gardeners encounter during the growing season. Possible solutions are listed.

- *Failure of tomatoes, peppers, and eggplant to set fruit (blossom drop).* If plants are growing well, this problem is frequently caused by adverse night temperature (below 60° F and above 70° F). Very seldom does heavy use of nitrogen fertilizers cause blossom drop, nor does sprinkler irrigation.
- *Blossom-end rot of tomatoes and peppers.* This is caused by a calcium deficiency that develops when soil moisture fluctuates (drought, heavy rains) or there is excessive nitrogen fertilizer. Proper irrigation, fertilization, and adequate mulch usually prevent it. Some varieties are more susceptible than others.
- *New leaves on cucumber plant suddenly wilt.* Leaves may show dead areas and fruit may be mottled. The most likely cause is cucumber mosaic virus, a common disease. Bacterial wilt and root rot are other possible causes of the wilting. A sudden rise in temperature or depleted soil moisture also causes wilting, but plants soon recover.
- *Lettuce and spinach go to seed.* This is normal for these crops under warm temperatures and long days. Planting in the spring and selecting the proper variety are remedies.
- *Kernels develop irregularly on sweet corn ears.* This may be caused by inadequate pollination. Planting sweet corn in blocks of several short rows rather than a single long row may help.
- *Snap bean flowers fail to develop.* High daytime temperature (above 90° F) is often the cause. Setting usually resumes when temperature drops.
- *Tomato fruits are rough and misshapen.* This is often associated with low temperatures (50°–60° F) while flowers are forming. The problem is worse on some varieties. The first fruit often are the most misshapen.
- *Cucumbers are off-shaped (e.g., crooked, nubbins).* This often occurs because of low soil moisture. Cool temperatures at the time flowers are developing can be a cause. Poor pollination because of lack of bees or low number or male flowers is also a possibility.

3

Vegetables Diseases

TOMATO

LEAF MOULD

Infection occurs mainly on the leaves, occasionally at the stalk end of the fruit. Leaf mould always appears first on the older mature leaves. The fungus enters through the stomata and the mycelium penetrates first between the leaf cells, afterwards into them. The dark conidiophores emerge in bunches through the stomata on the under-side of the leaf and bear numerous tiny elliptical conidia, mostly two-celled. They form a greyish bloom on the spots that turns a purplish brown with age. No conidia are produced on the upper surface of the leaf but yellow blotches appear corresponding to the mould spots on the under-side. Gradually as the disease progresses the leaves turn yellow, curl and finally shrivel completely, thus greatly weakening the plants.It takes the form of a black discolouration spreading outwards over the fruit from the end of the stalk but never covering the whole surface. The fruit may become lopsided and furrowed as a result of this attack. The fungus cannot attack the fruit directly but penetrates the stomata on the sepals or on the stalk and spreads through these into the young fruit or may prevent the fruit setting. Seed from infected fruit may carry the fungus and not only give rise to an infected seedling but act as a source of infection for neighbouring plants in the seed boxes. The seed coat containing the fungus is carried above ground on the cotyledons, conidia are produced on it and shed on to the surrounding seedlings.

As the fungus overwinters in the soil and on the interior fittings of the greenhouse the house must be disinfected after a tomato crop. This is most easily done by burning sulphur. One should not pass straight from an infected house into one where the crop is still healthy, nor take tools, baskets, etc., from one to the other. By such means the conidia can easily be introduced into the healthy house.It is of the utmost importance to prevent excessive atmospheric humidity. Therefore firing should be kept up at night throughout the summer and the house must be kept thoroughly ventilated. It is important to construct tomato houses so as to allow of plenty of side and end ventilation. An outbreak of leaf mould can be stopped altogether by heating and ventilating.

STEM ROT

The disease occurs mainly on the stems of medium-sized and old plants and on the fruit. It may also appear on the stems of quite small plants and sometimes causes dark spots on the leaves but this is exceptional. The rot on the stem forms a dark-brown area with a sharp boundary between it and the healthy tissue. Dotted over the brown part are,the tiny dark pycnidia. Infections can often be seen to be centred on wounds made by defoliation or other causes. In severe attacks the whole of the plant dies above the infected part. Infection occurs not only through cracks and wounds but also directly through the uninjured skin. The incubation period is 4 days or more.

The fruit is generally attacked only in one place but the rot may spread all over it. The spot usually appears when the fruit begins to turn red. It is then much the same colour as the rest of the skin but has a granular appearance due to the crowded pycnidia. The latter are usually very close together over the whole surface of the spot and eventually turn black so that the diseased area as a whole looks blackish. Under moist conditions pink globules of conidia ooze out of the pycnidia. Infection at the base of the fruit may cause it to fall.

ROOT ROT, FOOT ROT

The young plants become limp, the stem shrivels at soil level and they fall over. If the damage is caused by a species of *Phytophthora* or *Pythium* the shrivelled stem is greyish; if the damage is done by *Corticium solani* it is rather browner, but it is not always possible to distinguish the two forms of the disease without microscopic examination. Tomatoes are more often attacked by *Phytophthora* and *Pythium* than by *Corticium.* That of *P. cryptogea* Pethybr. and Laff. is 75° F. and that of *Corticium solani* is 65° F. None of these fungi causes appreciable damage at temperatures below 50° F.

All soil-borne fungus attacks can be prevented by sterilizing the soil by heat. *Pythium* and *Phytophthora* may also be controlled by formalin treatment of the soil before the seed is sown. Watering with Cheshunt compound can also be done to disinfect the soil before sowing and to check an outbreak of *Phytophthora* damping off in the early stages. Solutions of corrosive sublimate and saltpetre and proprietory preparations containing mercury may be used in the same way against *Corticium solani*. Drying the surface soil by raking helps to check *Phytophthora* but not *Corticium.* A few outbreaks have been reported but the disease is, no doubt, usually confused with other root rots. Affected plants wilt by day and recover at night. Usually their foliage is darker than normal and the under-side of the leaves often has a purplish tint. The stem is thin and hard and the leaves are small and pointed. When the plant is dug up or knocked out of its pot the main root is found to be dead with a curiously pointed tip. Usually the core of the root is brown and hard. New rootlets are formed higher up, only to be destroyed in their turn, and eventually adventitious roots arise from the stem above soil level.

Infection may occur in the seed box but it is seldom detected at this stage as the only symptom is blackening of one or more of the roots. In pot plants the main root is usually destroyed and the hard appearance of the tops may develop, but often the trouble is masked by profuse development of secondary roots. Usually toe rot is first noticed after the plants have been set in the border and made considerable growth. Then, when conditions favour rapid transpiration, the damaged root system is unable to supply the necessary water and daytime wilting sets in. Cold soils encourage the disease.

FOOT ROT (COLLAR ROT) OF OLDER PLANTS

The plants are attacked at soil level where a dark, sunken, decaying patch appears on the stem. The side roots are usually dead or brown before this occurs. The main root is also often brown and rotten, its soft parts decay and only the woody core remains. The disease may resemble stem rot but the lesions of that disease are bounded by healthy tissue above and below, whereas those made by *Corticium solani* start in the root and spread upwards so that they are bounded by healthy tissue at the top only.If outbreaks have occurred in previous crops the soil should be sterilized by heat or by corrosive sublimate treatment. When older plants are attacked by foot rot they can sometimes be saved by packing fresh soil round the base of the stem. If possible soil free from the fungus should be used. The stem then puts out new roots above the diseased part. To prevent the fungus from growing into the fresh soil it should be watered occasionally with solution of corrosive sublimate and saltpetre or with a proprietary compound of mercury. Neighbouring plants should also be watered with the same solution. Damage by this disease occurs particularly during cold springs and to weak plants, hence if the disease appears in a greenhouse the temperature should be raised. Care should be taken to see the plants make the normal rate of growth.

VERTICILLIUM WILT

Plants attacked by *Verticillium* wilt sometimes become stunted in growth because the stem ceases to elongate, thus causing the upper part of the plant to become rosette-like. At first the plants wilt only during the day and recover their turgidity during the night. The lower leaves develop light-orange spots, and gradually this discolouration spreads up the plant and the leaves first affected wither. Infected plants may die quickly or may remain in an unthrifty condition for a considerable period. In the latter case they may ultimately die or may recover from the attack.

The vascular bundles of the stem are stained brown. Usually the fruit is not attacked but occasionally the fungus causes a soft rot. Wilting of tomatoes on hot days may also be due to root trouble, including eelworm infestation.

BROWN ROOT ROT

For several years increasing attention has been paid to this disease in indoor tomato crops. Brown or brown-spotted roots appear in association with poorly developed tops. It has not been proved that the brown roots are the cause of the poor development above ground and the low yield of fruit, but circumstantial evidence makes this highly probable. There is often a small amount of browning on the roots even of plants that arc fruiting well. Brown roots are often seen also on melon, and the complex black root rot of strawberry seems to be a comparable condition.

If plants with healthy roots are planted in " sick " soil distinct brown roots may be found after 3-4 weeks. Even earlier microscopic examination discloses spots in which the root cells have yellow thickened walls while their contents are brown and granular. The boundary between these spots and healthy tissue is clearly denned. The brown spots or belts begin on the outer surface of the root, and gradually penetrate more deeply until they reach the vascular bundles, along which the brown discolouration spreads. When the roots are dug up one generally finds zones of brown, swollen and cracked cortex on the thicker roots alternating with zones of healthy cortex having the normal white or a light-brown colour. In more severe attacks the roots die until only a little tuft of weak fibrous roots remains at the base of the stem; the top is very backward, the whole plant thin and lanky and the fruiting very poor.

Many observations indicate that appearance of brown root rot is closely bound up with conditions of cultivation. In experiments at Cheshunt the disease was worst on overmanured, wet and compact soil and it was concluded that great variations in soil moisture should be avoided. Brown root rot may be very prevalent in a new tomato house, and with continued cultivation of tomatoes the brown discolouration may decrease with successive crops until it almost disappears as soil conditions improve. Replacing the soil of an affected greenhouse often helps greatly but the question as to whence this soil should be taken, which previous crop is most favourable to tomatoes, requires closer investigation.

SCLEROTINIA ROT

The plants wither, sometimes the whole plant dies suddenly, in other instances the lower leaves shrivel first. A rotten area, often covered by white mycelium, is to be found on the stem near soil level. Threads of the white mycelium may grow out into the surrounding soil. Wrapped up by the mycelium or enclosed by the stem cavity from which the pith has disappeared are large sclerotia, at first soft and white but ultimately hard with a black rind. Green tomatoes picked to ripen off may be destroyed by the same fungus if they come from infected plants and especially if they have come in contact with infected soil, or been placed on a bench where others had decayed before. They are rapidly converted into a rotten mass overgrown by dense white mycelium containing black sclerotia. *Sclerotinia* rot may be destructive both out of doors and in houses once the soil has become infected and if conditions are moist enough.

BUCKEYE ROT

Symptoms and cause. The two species of *Phytophthora* which cause foot rot and damping off of tomato seedlings can also attack the unripe fruit. This develops a greyish spot marked with concentric brown rings round the point of entry of the fungus. Intensive work on the physiological races of potato blight has led to the recognition of a race (A) widespread throughout the world and another closely related race (T) which has a particular capacity for infecting tomato leaves. Other races produce only larger or smaller spots on tomato leaves. In our experiments the yield of fruit has been increased and the number of diseased fruits reduced by spraying outdoor tomatoes with Bordeaux or Burgundy mixture as soon as blight is first seen on the potato crop, repeating the spray every 12 to 20 days thereafter. In experiments good results have been. obtained with copper lime dusts. Where outdoor tomatoes are picked to ripen indoors spraying definitely improves their keeping quality.

FUSARIUM FRUIT ROT

Symptoms. Infection begins at the stylar scar at the blossom end of the fruit as a small, sunken, soft spot that gradually increases in size and turns dark. From its surface break out pink conidial pustules of the *Fusarium,* often arranged in concentric rings. Ultimately the entire fruit becomes permeated, turns yellow, shrivels and dries up. The disease occurs on the outdoor crop. The attack is by the above-mentioned species of *Fusarium* which penetrates the fruit through the tiny wound left by the fall of the style. It cannot penetrate the unbroken skin but on picked fruit it easily enters by the stalk scar and may cause serious loss of outdoor tomatoes ripening in store. The fungus needs a high temperature and high atmospheric humidity for active development.

The disease almost invariably occurs on the lower trusses of fruit. The fungi concerned inhabit the soil and attack fruit in contact with the ground, or those on to which they have been splashed during watering.Buckeye rot is well known in many countries where tomatoes are grown and has caused great loss from time to time.

VIRUS DISEASES OF TOMATO

Common Tomato Mosaic

Symptoms vary from ill-defined, light-green spots, mainly between the veins of the leaflets, to quite a bright, clearly defined, light and dark-green mottle. The surface of the leaves may also be slightly uneven or even curled. Tomato variety, nutrition and the time of year all affect the expression of symptoms; thus, during the sunny warm days of summer the mottling is most obvious, and during spring and autumn the curling becomes apparent. At low temperatures the symptoms may be almost entirely lacking but the plants remain infected and infectious to others. Flowers of infected plants may be deformed, sterile and fall off. There may be slight streaking of the stems and pale, slowly ripening spots on the fruit. Infection from tomato plants. The invisible infectious matter, or virus, occurs in the sap of almost all the cells of an infected tomato plant, including the cells of its numerous hairs.

These are easily broken and it is a familiar experience that one cannot handle a tomato plant without getting sap on one's hands, tools and clothes. If one touches an infected plant and then a healthy one a mere trace of sap transferred from one to the other is enough to transmit the infection. Hands, knives and other tools are the most important agents of spread. Aphides do not spread this disease. The incubation period is from 5 to 20 days or 28 days in our experiments out of doors. It persists for a very long time in dead plants and purified virus preparations can remain viable in the laboratory for 5 years. The virus is not killed until heated for 10 minutes at 194° F. or until treated for 24 hours with 60 per cent alcohol, but it is sensitive to soap. Probably tomato sap dried on clothes can remain infectious for a long time. Virus liberated from a plant into the soil can remain infectious for a long time if the soil is moist; freezing and drying the soil renders it non-infective.

Infection from other plants and products. Common tomato mosaic is caused by the same virus as common tobacco mosaic, which is common in all tobacco-growing countries. Fermented, commercialj smoking or chewing tobacco may retain the virus in an infectious condition. Hence chewing tobacco is a common source of infection in commercial tobacco production. Besides species of *Nicotiana* this virus can infect most plants belonging to the family *Solanaceae* and also species in other families, notably asclepias, snapdragon, larkspur, zinnia, beans, spinach and possibly celery. Potato, black nightshade and woody nightshade (bittersweet) show no symptoms, but they are carriers of infection.

From these hosts the virus can be returned to tomato.

The nightshades afford a means of carrying it over winter as also do susceptible greenhouse ornamentals like *Petunia, Solanum* and others. There seems little danger to the tomato crop from mosaic-infected potatoes though there is evidence of one such a transmission out of doors.

It may be supposed that the virus multiplies more luxuriantly in vigourously growing plants. The statement that a crop can be " manured out of mosaic " only means that suitable manuring may overcome the consequences of infection; the virus in the plants is not destroyed. When nitrogen is very deficient the mottle becomes invisible; if nitrogen is then supplied it reappears. The influences of light and temperature have been mentioned above. Attacks are often found to start near doors and ventilators, which can scarcely be due to draughts but rather to infection from outside the house.

SPOTTED WILT

Then a large number of tiny bronze-coloured spots appear either in rings or scattered over the whole leaf surface. Growth stops for a time so that infected plants appear stunted. Young seedlings may be killed but more often the plant resumes growth after a time, but its leaves become distorted and have a rather conspicuous yellowish mosaic mottling. The fruit may be unaffected or it may bear pale-red, yellow or almost white markings, often in concentric rings. The virus is sap-transmissible but not to the same degree as tomato mosaic. It is not spread by aphides but only by thrips. Curiously enough it is only in the larval stage that thrips can take up the virus from an infected plant, but once they have done so they remain infective throughout their life. *Lycopersicum* Virus 3 is not very resistant, it remains viable only a few hours in expressed tomato sap, is inactivated by drying, by oxidizing agents and by exposure for 10 minutes to a temperature of 108° F. It has a very wide host range and commonly infects many popular greenhouse perennials which then form reservoirs of infection for the next tomato crop. Most of these show little sign of infection or bear only a few rather inconspicuous ring marks on their leaves.

TOMATO YELLOW LEAF

It starts as a well-marked mosaic mottle on the youngest leaves. The older leaves turn yellow, except along the main veins which remain dark-green. As the plant grows the yellow spreads until only the topmost leaves remain green. The whole picture resembles nitrogen starvation. Some evidence of seed transmission has been obtained. In properties the virus closely resembles that of common tomato mosaic. The aster yellows virus can also infect tomato and causes a certain amount of yellowing and a bushy type of growth, as it stimulates development of leafy buds and shoots in the axils of the leaflets as well.

AUCUBA MOSAIC

This disease is caused by a strain of the common tomato mosaic virus and must not be confused with aucuba mosaic of potato, caused by quite a different virus. Bright yellow spots often appear on the leaves of tomato plants infected with common mosaic. If these are punched out and used for inoculation the aucuba mosaic disease is produced. It is also sometimes found by itself in commercial tomato houses. About 5 days after inoculation the youngest leaves curl down and their surface becomes rough, wrinkled or corrugated. By the seventh or eighth day yellow spots appear and become gradually larger, until in extreme cases the whole leaf surface is yellow, with a few raised islands of dark-green standing out on it like blisters. More often the yellow and green areas are of about equal extent and intimately intermingled. Plants are not killed by the virus but their growth is checked and the fruit may be mottled. Properties of the virus are the same as for common tomato mosaic. There is some evidence that seed-borne infection can take place in a very small number of seedlings, perhaps by virus dried on the seed coat.

TOMATO DISTORTING OR ENATION VIRUS

This is another strain of the common tomato mosaic differing in that, in addition to causing mottling of the leaves, it stimulates the plant to excessive and unnatural growth. The leaflets may be increased so that a fern-like effect is produced with very many small leaflets crowded together, gradually diminishing in size and ending in a corkscrew-like tendril. In some cases small outgrowths (enations) are formed on the under-side of the midribs. There is a tendency to excessive growth and production of giant plants crowded with foliage.

BUSHY STUNT

This is a distinct and apparently rather uncommon virus which is easily transmitted by sap but has no known insect vector. In seedlings it causes yellow and purple colouration of the leaves, yellow spotting and death of the young leaves and often a sunken brown lesion at soil level, causing the seedlings to fall over. In older plants it causes cessation of growth by the terminal shoot and stimulation of the side shoots, hence the name " bushy stunt." The lower leaves turn yellow or purple and often there are concentric rings of bright yellow or purple, or purple lines along the veins of a yellow leaf. Ultimately the lower leaves shrivel and drop off. This virus is particularly interesting because it has been purified and obtained in the form of minute cubical crystals.

MISSHAPEN FRUIT

The tendency to production of ridged and misshapen fruit is an inherited one, as shown by Pape. It is also, however, influenced by conditions of cultivation. Thus, Kondine Red is a variety rather prone to bear ridged fruit but it only does so when the plants are set very far apart and richly manured. To avoid this disfigurement seed should be saved only from carefully selected plants with well-shaped fruit. The seed crop should be grown under conditions conducive to misshapen fruit so that any inherited tendency to do so becomes evident. In normal commercial fruit production spacing and manuring should be that found by experience to suit the variety grown.

BLOSSOM-END ROT

The most striking symptom is a dark, more or less sunken, area at the blossom or style end of the fruit, i.e. farthest from the stalk. The trouble begins as a water-soaked spot at or near the base of the style, or the whole of that end of the fruit may look as though bruised. The affected part soon becomes leathery and usually turns black. As the rest of the fruit continues to swell the blossom end becomes slightly sunken or remains abnormally flattened.

Various bacteria and fungi are commonly to be found in the affected tissues and have from time to time been described as the cause of blossom-end rot. They are, however, purely secondary invaders, living in tissue already damaged by a physiological disturbance. Especially if the surface of the fruit remains moist these secondary organisms may enter and set up an extensive rot. Tomato varieties in which the style remains attached for a long time seem particularly susceptible to blossom-end rot, perhaps because fungi and bacteria become established on the dead style and are hence ready to invade even a small spot of true physiological blossom-end rot.

The most important preventive measure is regular watering to avoid alternate drying out and soaking of the soil.

It may be worth while to pick off the dead corolla and style on varieties that retain them long after the fruit has set. Otherwise fungi or bacteria established in the dead tissue may set up secondary rots following blossom-end rot. The variety Comet seems particularly susceptible to this trouble.

HOLLOW FRUIT

Fruits divided by a single septum into two compartments are particularly prone to develop empty spaces between the septum and the fruit wall. Such fruit when shaken often makes a rustling or rattling sound. A larger or smaller part of the seed and flesh inside may be black and dead. This unfortunate phenomenon, which appears especially towards the end of the summer, is not due to attack by any parasite. It is due especially to a period of drought during the swelling of the fruit but other growing conditions may possibly cause it. In such circumstances when there is a heavy crop only part of it will contain normal contents.

GREENBACK

Greenback appears as a hard green or yellow area at the top of the fruit, surrounding the stalk. When ripening starts this part does not turn red like the rest but remains green and then slowly goes yellow or orange. It never becomes red and soft like the rest of the flesh. Greenback, like blotchy ripening, is due to a deficiency of potassium, especially in fruit exposed to strong sunlight. Tomatoes forced to ripen at a temperature exceeding 90° F. do not turn red but yellow. Lycopin, the red colouring matter, is not formed at such high temperatures. Greenback always occurs on the uppermost part of the fruit as it hangs on the truss, which is, of course, the part most exposed to the sun.

When defoliating tomatoes one should be careful to leave enough leaves to shade the trusses during the hottest part of the day. If the foliage is scanty, side shoots should be allowed to grow out and produce a couple of leaves before being stopped. Glasshouses should be shaded during the hottest periods. Care should be also be taken to supply enough potassium and water. Kondine Red and Tuckwood are very prone to greenback, Selandia much less so.

BLOTCHY RIPENING

In blotchy ripening larger or smaller parts of the fruit surface do not change colour with the rest as the tomato ripens. When it begins to turn red they remain green for a time but gradually turn yellow, develop a waxy appearance and a turnip-like flavour. The vascular bundles in the blotches are often dark and show faintly through the skin as dark lines. The blotches may occur on any part of the fruit, a feature which distinguishes them from greenback which occurs only round the stalk. Blotchy ripening is seen mainly in June and July. Tomato varieties differ greatly in susceptibility to it; Kondine Red is one of the most susceptible. Blotchy ripening is due to cultural factors or to diseased roots. In manurial experiments at Cheshunt it has been demonstrated that the greatest number of fruits with blotchy ripening occurred on unmanured plots and on plots deficient in potassium and nitrogen.

If tomato roots are diseased, if they receive too much water or too little, they cannot absorb enough nutrients and this may lead to blotchy ripening. The dampness of the subsoil should therefore be taken into account. It may well happen that the top 6-9 inches of soil are fairly moist but that the deeper levels are too dry. Tomatoes seem to develop most blotchy ripening in heavily limed soil. In Danish nurseries blotchy ripening is often prevalent where there is a large amount of phosphoric acid in the soil. If a heavy dressing of manure is given the roots may be damaged and thus less able to absorb nutrients so that the plants may starve on over-manured soil. Blotchy ripening is usually worst in new houses where growth is unusually vigourous. In summers with plenty of sunshine less potassium and more nitrogen should be supplied than in dull, cloudy seasons.

CRACKED FRUIT

Cracking of tomato fruit occurs in two chief forms, single large cracks and many small ones; both types occur mostly on the top half of the fruit. The large cracks happen especially when a dry period is suddenly followed by a damp one. They are therefore usually found in the outdoor crop, where the grower has least control over the moisture relations of the plant. Cracking may also result from a sudden acceleration of growth for other reasons, as after manuring. The skin makes poorest growth under cool conditions and then has difficulty in keeping pace with rapid expansion of the flesh.

The numerous small cracks on tomato fruit occur especially in greenhouses where there are great fluctuations in atmospheric humidity. Growers should therefore be careful not to stop night heating too early It is probably the increased root development that counteracts the effect of a fluctuating water supply. Potassium permanganate is usually used for watering, etc.

ROLLING OF THE LEAVES

Tomato leaflets sometimes roll up so much that the upper surface becomes completely overarched by the upcurled margins which may meet above the midrib of the leaflet. Such leaves feel very hard and stiff. Varieties and individual plants may differ somewhat in the degree of leaf rolling. Tomatoes, perhaps because they are more extensively stripped than the glasshouse crop. The rolling of tomato leaves has no connection with the leaf roll disease of potatoes, which is due to virus infection.

Rolling up of the top leaves may occur on young, actively growing plants. It occurs mainly when the crop has been heavily manured so that there is a high concentration of nutrient elements. Such plants have dark-green foliage and the top may have a violet tint somewhat reminiscent of spotted wilt.

It occurs mainly in spring, particularly on tomatoes but also on other greenhouse plants such as pelargoniums and early forced potatoes. The outer cells become gorged with water, swell and burst the skin. The swellings then have a white powdery appearance but gradually turn brownish as the broken tissue falls away. Affected leaves have a tendency to roll.

Edema occurs when the roots are absorbing more water than the leaves are able to transpire. A warm, moist soil increases the rate of water absorption but damp air, poor ventilation and shade check transpiration from the leaves. The condition may also become serious when, after a cool period with slow growth, the temperature rises and supply of water and nutrients to the foliage is suddenly increased. Such plants have relatively few leaves unable to cope with the large increased water supply resulting from the altered conditions. To check oedema the soil should be kept fairly dry and there should be thorough ventilation.

This condition arises in plants that have became much more prevalent after the adoption received too much water or excessive nitrogen of petroleum oil sprays against red spider. The or have experienced a sudden change in con-film of oil checks transpiration from the leaves, ditions when planted out. The grower often finds a few plants which times been noticed some time after severely early attract notice by remaining short because pot-bound plants have been planted out. Once the roots have grown out into the new than in normal plants. These should be rogued soil they are able to absorb more water than out before the crop is planted out as such can be dealt with by the rather stunted tops, plants are partially or completely unfruitful.

Some time after planting out the stems growth. As a rule there are only a few of these may become soft and flaccid; in bad cases the in each batch of seedlings. Like the short-plants fall over without any actual rot taking jointed plants they are practically sterile, place. When such a stem is cut open the pith These varieties seem to be due to inherited is found to be wholly or partly wanting. Plants defects. They occur scattered amongst normal that have fallen over in this way may remain plants and should be weeded out as early as green for a long time but make no growth. possible.

POTATO

POTATO BLIGHT

The attack on potato leaves starts as a rule at the edge or tip of a leaflet as a light-green spot, which quickly becomes brown and limp, and later blackish-brown and dry. Under damp conditions the brown spots have a marginal zone of fine white mould, composed of conidiophores and conidia of the fungus, which burst through the skin of the leaf 2-3 days after infection. There are at first a few scattered spots, usually on the lower leaves; but others quickly follow and the whole potato haulm may be destroyed in about a week. The diseased leaves have a characteristic sickly sweet smell which can often be detected at some distance from a diseased crop. Brown streaks develop too on the stems and even the flowers and berries may be attacked.

Potato blight on under-side of the leaf. Tubers attacked by potato blight show leaden or dull purplish-grey, slightly sunken spots or irregular blotches in the skin. When they are cut through, rusty-brown streaks are seen growing from the skin into the flesh. In time the whole tissue becomes destroyed by a rather dry brown rot. Commonly, however, *P. infestans* is quickly followed by secondary fungi, especially species of *Fusarium* and bacteria, which obscure the symptoms of blight and may cause an evil-smelling soft wet rot to supervene. In all its stages the blight fungus needs oxygen and it cannot penetrate rapidly into the flesh by itself. The secondary invaders feed on the starch grains that the blight fungus has only superficially attacked. *P. infestans* conidia may be formed on diseased tubers.

The most important source of infection every summer is certainly slightly infected seed tubers. These may give rise to

1. Quite a normal sprout,
2. Infected sprouts which soon die or
3. Infected sprouts which come above ground, bear conidia and thus infect the lower leaves of the crop.

The potato blight fungus can also grow out from diseased tubers and small dead shoots into the soil where it lives for a time, and may produce conidia which are splashed on to the leaves by rain. The appearance of the fungus in a crop from diseased seed tubers has been doubted but was seen by de Bary. Severely attacked tubers are readily destroyed by bacteria and millipedes, and sprout feebly or not at all. It is only those tubers which are slightly infected which are able, when planted, to produce shoots that come above ground. In June and July spread of infection from these diseased sprouts to lower leaves of the crop proceeds unnoticed. These centres of infection form starting points for the incredibly rapid blazing up of the disease which occurs when conditions become favourable to it in July and August.

It may be supposed that it will break out at many places at the same time if the conidia are carried a short distance by the wind. It may be of importance whether the early potatoes lie on the windward or leeward side of the late varieties. Sometimes one may notice that a hedge has afforded a certain amount of shelter and that blight has started opposite a gateway or other gap in the hedge. Small potato patches in town gardens become attacked at the same time as fields in the country. In experiments with cultures in sterilized soil or manure, where the blight fungus encounters no competition from other organisms, it has survived the winter out of doors. The possibility of it overwintering in ordinary field or garden soil is very small but it must be remembered that there are always some groundkeeper tubers not killed by frost and the fungus may naturally survive the winter in some of those. Heaps of chats discarded when the crop was graded or left on the site of an old pit are another possible source of infection.

If the potato tops are killed early by blight and the tubers are left to ripen in the soil a very small crop is obtained but the tubers do not become infected, provided they were covered by a good thick layer of soil. If blight attack comes late, retarded by weather conditions or by spraying, so that the tops are still green and bear fresh blight spots when the crop is lifted, there is great danger of the tubers becoming infected, especially in wet weather. Liability of the tubers to become infected depends partly on the variety, partly on the soil. Haulm and soil remain infectious for 10, possibly 20, days after the haulm has died. Haulm cut down by frost is at once freed from infection.

ALTERNARIA BLIGHT

This leaf spot disease is known and feared under the name " early blight " because it tends to appear earlier than true potato blight, called there " late blight." It is due to the mould *Alternaria solani,* formerly called *Macrosporium solani* Ell. & Mart., which causes leaf spotting also on tomato and egg plant.

More or less circular brown or blackish spots appear on the potato leaves from July onwards. They are limited by the leaf veins though in severe outbreaks several spots may become united. Under a lens the spots are seen to be marked with darker, irregularly concentric lines, hence the name " target spot" sometimes applied to them. In damp weather a fine olive-brown mould may appear on the surface. Some varieties show considerable resistance to the disease and only very small brown spots develop on their leaves.

This is the form of *Alternaria* blight usually seen in Britain, where, indeed, the disease is regarded as of no economic importance and where conidia are seldom to be found on the spots. It seems clear, in fact, that not all the " target spot " on potato leaves in this country is attributable to *Alternaria* infection. Very early spraying with Bordeaux mixture prevents an outbreak of *Alternaria* blight; according to one investigator Burgundy mixture is even more effective. Where the disease has been prevalent care should be taken that the tubers do not become infected by spores from the haulm at lifting, and to store them in a dry cool place. The optimum temperature for *Alternaria* rot of the tubers is 60° F. and its growth ceases at 41° F. Alpha, Bravo, King Edward, Majestic, Kerr's Pink, Edzell Blue and Up-to-Date are all susceptible; Juli is fairly resistant and Golden Wonder more so.

This fungus causes a minor leaf spot disease which in some parts of Europe causes premature withering of potato haulm in severe seasons. The leaf spots appear from July onwards; they are at first yellow, then brown on the upper surface, and are tinged greyish-purple with the conidia on the under-side of the leaf. The routine sprays against potato blight control this disease as well.

POTATO WART DISEASE

Wart disease increases rapidly on land where potatoes are frequently grown, as in gardens md allotments, and in a space of 2-3 years it may attain such proportions as to ruin the crop. Direct losses of crop from the disease on an agricultural scale have been small but stringent quarantine regulations against it hamper export of produce from infected districts and discourage the cultivation of susceptible varieties. Hence the regulations which aim at excluding wart disease from countries in which it does not yet occur cannot be held to be unreasonable. On several occasions the disease has been brought in with small consignments of potatoes among immigrants' luggage or by small-scale commercial traffic across a frontier.

Only plants of the nightshade family, *Solanaceae* can be attacked by this fungus. Tomatoes develop swellings on roots and stem, and under experimental conditions small warts on the leaves also. Henbane and a number of species of *Solanum* develop small swellings on the roots and eventually small galls on the stems. It has not proved possible to infect thorn apple *(Datura stramonium],* deadly nightshade *(Atropa belladonna)* or any kind of tobacco.

In a wet season warts may be found on the stems and lowest leaves of potatoes, and even on the inflorescence, but as a rule the attack is confined to the tubers and is first noticed at digging. The stolons may be affected but the roots are never attacked. As a rule growth of the haulm is normal and unchecked. The warts on the tubers start as tiny white granular swellings growing out from the eyes. They quickly enlarge into cauliflower-like structures which may be larger than the tuber itself. Above ground they are light green, below ground at first white, then brown.

Old warts decay and turn black and even at the normal lifting time many will be partially rotten, or break off and decay in the soil, leaving only a small scar on the tuber. Occasionally new small outgrowths may develop in the clamp around the scars, giving a disease picture like that of powdery scab. Powdery scab infection of the sprouts occasionally produces swellings simulating wart disease, as may also irregular second-growth, covered with scab. Sometimes microscopic examination is necessary to distinguish between these diseases. The swellings of crown gall also resemble those of wart disease but the former is known on potato only from artificial inoculations.

The wart disease swellings contain a vast number of minute, brown, thick-walled resting sporangia which become liberated into the soil as the tissues decay. When they germinate a number of extremely minute, mobile swarm spores emerge and swim about until they meet a young potato tuber. They then penetrate the skin of an eye and settle down inside the potato cells.

Here in a fortnight clusters of from i to 7 summer sporangia are formed which liberate hundreds more swarm spores. These are partly responsible for multiplying the number of infected cells on and around the developing wart, partly for infecting other tubers. Second generations of summer sporangia form during the summer, and their development requires 10-20 days. They do not overwinter but are replaced in the mature wart by the thick-walled winter sporangia, formed after a kind of sexual process. These not only survive the winter but remain alive in the soil for very many years; they are not destroyed even when an infected potato is eaten and digested by an animal.

The formation of warts is due to the potato cells being stimulated to abnormal growth and multiplication by the fungus cells which invade them. The wart disease organism is one of the most primitive of fungi and never develops any kind of mycelium.

METHODS OF SPREAD AND REQUIREMENTS FOR INFECTION

The most important and most dangerous means by which infection spreads is the transport of seed potatoes, not only diseased tubers but healthy potatoes from infected land, which may carry resting sporangia in the soil adhering to them. Potato peelings in kitchen waste added to compost heaps or manure and dung of animals fed on raw potatoes may also carry infection. The fungus may be carried from infected land in soil on carts, tools, boots, and the feet of domestic animals and poultry. Soil particles transported by storms, or floods, or by crows and other wild birds may also spread the disease. Resting sporangia can germinate over a wide temperature range (40 to 86° F.) but best at 66-68° F. Presence of oxygen and nitrates encourages germination. Invasion of the host cells by the swarm spores also occurs over a wide temperature range (45-86° F.), best at 60-68° F., and oxygen is needed for this process. Soil reaction seems to have little effect on infection. It occurs over the whole range between pH 3.9 and 7.5, even pH 5 is said to be quite favourable. Soils more acid than these are seldom cultivated and one can feel no assurance that infection will not occur on soils more alkaline than 7.5. No essential difference has been found between infection on clay and sandy soils. Extreme drought may completely prevent outbreak of potato wart disease. On the other hand the warts become particularly large with a heavy fall of rain. Plants that mature early or form no tubers owing to virus disease, etc., may escape attack, as may very early varieties. Resting sporangia in the soil survive longest when the ground is under permanent grass or is left unbroken for other reasons. After a severe outbreak infection has been proved to persist in the soil for 12-16 years.

Wart disease is easily and completely controlled by growing immune varieties. No potatoes from an infected crop may be sold for planting and no tubers visibly affected with wart disease may be sold for any purpose. Only approved immune varieties of potato may be grown on land where wart disease has occurred at any time. In a scheduled Protected Area the only potatoes allowed to be planted are tubers from a crop certified to conform to official standards of purity and health and to have been grown on land free from wart disease.

It is also forbidden to remove rubbish or compost from such places or to dispose of manure. Export of tubers from boroughs where infected areas occur is also prohibited. There is no practicable method of destroying wart disease resting sporangia in a field soil. On a very small scale soil can be disinfected by steaming; baking the soil is not effective.

Treatment of soil with 5 per cent formalin for 24 hours also kills the sporangia. Incorporation of sulphur in the soil also has some effect but only in large quantities which make the soil sour and are unreasonably expensive. Implements used on infected land may be disinfected by dipping in 2 per cent formalin solution, but disinfection of potato " seed " has not proved effective. Cooking of potatoes used as fodder helps to prevent contamination of manure and is, in any case, required by the Wart Disease Order as far as infected tubers are concerned.

COMMON SCAB

The disease called common scab of potato is caused by the ray fungus *Actinomyces scabies,* which also attacks beet, carrot, radish, swede and turnip. Care is needed to distinguish between it and powdery scab. The ray fungi are generally regarded as being somewhat intermediate in structure between bacteria and fungi. Several hundred species occur in soils, where they play a large part in the breakdown of decaying organic matter; others live in water and a few cause disease in animals. Potato scab appears as rough, scabby spots on the otherwise smooth skin of the tubers. The spots may grow together and the whole surface of the tuber may be covered by them. If careful examination is made when the layer of cork begins to form in the skin the first signs of disease will be detected as small grey-brown spots under the lenticels where the fungus is stimulating the living cells to production of additional layers of cork. The outer cork is burst open and the new cork follows suit, producing the scabby outer surface of the lesion. The more prone the variety is to cork formation the scabbier it will be.

Roots and stolons of potato may be slightly attacked. Tubers that protrude above ground are not affected, nor do scabs increase in the clamp. Tubers are the most readily infected; if they have a long growing period, the period during which they become scabby is increased, hence, perhaps, the various reports that " spraying induces scab." A strong earthy smell due to the presence of the actinomycete is peculiar to common scab; powdery scabs have no such smell.

The fungus may overwinter either in the soil or in the scabs on tubers. It can survive on cellulose and similar plant products in the soil for years at a time. Several instances have been reported in which disinfected tubers planted in freshly broken forest soil produced scabby crops. On the other hand an interval of several years between potato crops has sometimes led to a striking reduction in the amount of scab.

The fungus reproduces simply by its fine hyphae breaking up to form spores, easily disseminated in soil and water.

Manure may be regarded as infected if it comes from farms where scabby potatoes are used raw as fodder. Infection may also be spread by compost and, of course, by planting infected tubers.

VARIETAL SUSCEPTIBILITY

No potato variety completely resistant to common scab is known and on soil favourable to the fungus all varieties are attacked to about the same extent. Early varieties tend to escape infection, probably because the cork layer in their skin is formed before the soil becomes warm enough to suit the fungus. Main crop varieties do show some differences in susceptibility, probably associated in part with the size and texture of their lenticels. Breeding for scab resistance is in progress in Minnesota where the possibility of physiological races of the fungus is also under investigation.

Disinfection of the tubers is effective in destroying the scab fungus on them but is only of value when they are to be planted on scab-free soil, or in soil treated with sulphur or otherwise disinfected. Digging in grass cuttings may be effective in gardens or allotments provided they are available in sufficient quantity. Up to a stone per square yard has been recommended. On a field scale ploughing in rye or other green manure in the spring may be helpful.

Consistent use of sulphate of ammonia instead of chili saltpetre will gradually reduce the amount of scab, but one must be careful that the soil is not made too acid for other crops that need lime. Combined use of dung and sulphate of ammonia has proved helpful. The proprietary preparation Bayer P. at the rate of about *I* oz. per square yard or 54 Ib. per acre, dug well in before the potatoes are planted, may greatly reduce both common scab and stem canker. It works very well when used with sulphate of ammonia.

Liming may be necessary because of other crops in the rotation, but it is well to have the soil analysed and to apply only the lime requirement. Where it is important to obtain a crop of smooth tubers a piece of ground may be kept unlimed for potato production. Continuous cropping with potatoes is liable, however, to lead to trouble from other diseases and especially from root eelworm.

POWDERY SCAB

Powdery scab, sometimes known also as corky scab, first appears on the skin of potato tubers as pale little pimples. Either before or after lifting, the skin over the pimple splits and falls off, exposing a snuff-brown powder composed of the spore balls of the fungus. For a considerable time the individual powdery scab remains as a shallow crater-like hollow, filled with brown powder and fringed by the loose edges of burst skin. Under good cultural conditions the scab seldom exceeds, but in potatoes grown in wet land or which have made considerable second growth, large open scabs and cankers may be formed, especially on the ends and protuberances of the tuber. In such extreme forms of the disease microscopic examination may be needed for confident diagnosis.

Corticium solani may be found in the wounds and various other secondary organisms may enter and set up a rot. In exceptional circumstances the fungus may continue its activity on tubers in the clamp and stimulate them to production of irregular tumour-like outgrowths closely resembling those of wart disease. *Spongospora subterranea* also attacks the roots of potatoes, and less often of tomato, causing irregular white nodules.

It is doubtful if these cause any damage to the plant but they contain spores which are liberated into the soil when the roots decay and thus add to the reservoir of infection in the ground. *S. subterranea* is one of the most primitive fungi, a relative of the fungus which causes club root in cabbages. Its spores, which are stuck together in peculiar sponge-like balls, can remain alive in the soil for 4-5 years at least and are easily spread to new localities in soil on tubers or by adhering to roots of transplants.

When they germinate they liberate swarm spores which penetrate either the roots through root hairs or the tubers through lenticels or small wounds. Cultural conditions. Powdery scab is found on both sides of the Atlantic and in the potato-growing districts, but is only of importance where there is a heavy rainfall and on ill-drained soils. In dry soils a reasonably clean crop may be lifted even though grown from infected tubers. No varieties are immune from the disease but there is some difference in susceptibility, said to be correlated with the thickness of the cork layer in the skin of the tuber. Liming is said to make the trouble worse, and application of sulphur at the rate of 3-6 cwt. per acre checks it.

Where the disease is prevalent, soil drainage should be made as efficient as possible and the rotation should be arranged to leave 4-5 years between potato or tomato crops. Soil disinfection is impracticable in the field. Diseased tubers should not be returned to the land but may be cooked and used for fodder. Manure from animals fed on raw infected potatoes will contain undamaged spores and should not be used for potato fields, nor should potato peelings and similar kitchen refuse be put in the compost heap. Disinfection may be practised where infected tubers have to be used on land free from the disease. They may be dipped for 3 hours in formalin solution, i pint of formalin in 30 gallons water, or 1: 2,000 corrosive sublimate solution. Moistening the tubers and rolling them in powdered sulphur may also be effective. Spores of closed scabs still in the pimple stage may not be killed by any of these methods.

Secondary scab injury. Both in common and powdery scab, especially when associated with second growth, the damage is often increased by the activity of a number of insects and other pests, such as the larvae of *Sciara* flies, wood lice, and millipedes which penetrate the outer layers of the tuber flesh by way of the scabs and cracks and convert it into a dry crumbling crust. The stem eelworm *Anguillulina dipsaci* produces extreme crust-like rotting of this kind. Secondary fungi may also play a part; even dry rot often starts at a powdery scab lesion. When storage conditions are too damp, or too warm, soft-rotting bacteria obtain easy access to the damaged tissue and the whole tuber decays.

BLACK SCURF AND STEM CANKER

This, one of the commonest and most cosmopolitan of fungi, was first recognized in connection with potatoes, with which it is almost invariably found. But it also attacks many other plants, cabbage, tomato, wheat, lettuce, asters, though sometimes in the form of a specialized strain more or less adapted to a particular host, and has been found on many weeds such as buttercup, shepherd's purse, field thistle, plantain, etc. It can also live freely in the soil.

More recent studies have made it possible to sub-divide this very large genus into a number of smaller more clearly denned genera. When this is done C. *solani* falls in the genus *Pellicu-laria* and some authors think it identical with a fungus described earlier under the name *Hypochnus filamentosa* Pat., hence it has been re-christened *Pellicularia filamentosa.* This name has not yet, however, been generally accepted. On potato tubers one often finds hard scurfy patches to which the soil adheres. When the tubers are washed these patches show up as dark-brown or black spots and lumps. They are the sclerotia of C. *solaniy* sometimes described as "The dirt that won't wash off." They are attached to the skin by very fine hyphae and can be scratched off with the finger nail, leaving the uninjured skin of the tuber beneath. Apart from making ware potatoes a little unsightly this phase of the fungus's life history does no immediate damage.

Occasionally it has been found to attack tubers through the lenticels before they are lifted, causing small shallow brown pits rather like those of pit rot. When "seed" tubers carrying C. *solani* sclerotia are planted, however, the fungus becomes active again and many cause more or less serious injury to the sprouts. It may be found on the young sprouts even in the chitting boxes but most damage is done after planting.

Then sunken brown spots appear on the sprouts below soil level. Or the tip of the shoot may be killed before it emerges above ground; it then puts out a side shoot which may escape or may also be killed. In the worst cases groups of such dead shoots are found with the result that blanks occur in the potato drill.

Later in the summer the fungus causes dry, brown, sunken wounds on the underground parts of the stems. If these are deep and girdle the stems the haulm may be weakened and the leaves roll, somewhat as in primary leaf roll though less stiffly. When the starch cannot reach the tubers, green aerial tubers are produced in the axils of the leaves. Brown lesions also appear on the stolons, which may be killed. The plant then forms merely a cluster of small angular tubers round the base of the haulm.

The roots too may be attacked, especially the fine absorbing rootlets.In all these brown sunken lesions only the mycelium of the fungus, or occasionally *Rhizoctonia* sclerotia, can be found. But at the base of vigourous haulms where the air remains moist the fungus covers the stems just above soil level with a white powdery film, sometimes extending up them to a height of 6 inches. This film consists of a layer of basidia and basidiospores. No harm seems to be done to the plants by this sporing stage of the fungus but it offers special possibility for its spread by means of wind-blown spores. The effect of an attack on yield of tubers varies greatly from practically nothing to severe losses where the fungus has caused extensive gaps in the drills.

Conditions of growth. Potato sprouts may be infected partly from the see"d tubers, partly from the soil. The fungus seems to thrive in all cultivated soils, is not much affected by soil reaction and can grow over a wide range of temperatures.

The rate at which the tubers sprout determines the severity of the disease in spring because the damage is greater when potatoes are planted early in cold soil; nevertheless early planting generally pays! As a rule lavish use of dung and artificial manures helps the crop.

VERTICILLIUM WILT

The symptoms of *Verticillium* wilt in potato usually develop only after the haulm has attained a normal size, though in extreme cases it may be stunted, with all the leaves rolled, and yields a negligible crop. More often, from July onwards, the leaves become limp and turn yellow at the tips and edges. Often one side of the leaf turns yellow and dies. Finally the leaves die and hang down brown and shrivelled against the stem, so that the whole plant gradually dries from below upwards. Characteristic of the disease is the brown stain in the woody strands, often extending throughout the whole haulm. In autumn a delicate velvety white mould of conidiophores may appear on the base of the dying plants, and finally the stem may be blackened by the minute resting bodies of the fungus.

Tubers on infected plants are usually also infected from the stolons, and if one slices them at the heel end the vascular bundles will be found to be brown, often far into the flesh. Brown vascular bundles occur also in *Fusarium* wilt, black-leg and other diseases, including leaf roll, and some forms of slight frost damage. There is no natural separating layer between the tuber and the stolon so that when the crop is lifted a small wound is inevitably made at the heel end which takes a little time to heal. Before the cause of potato leaf roll was understood it was frequently confused with *Verticillium* wilt and there was consequently a tendency to exaggerate the prevalence and importance of the latter disease.

EFFECT OF ENVIRONMENT

Verticillium wilt is widespread on both sides of the Atlantic but in warm climates it tends to be superseded by *Fusarium* wilt. High temperatures check the growth of the fungus but the drought and excessive transpiration accompanying them usually kill all the infected potato plants. Experiments have indicated that heavy nitrogenous manuring increases the susceptibility of the crop; soil reaction has little effect.

Provided a normal rotation of crops is practised the disease is unlikely to become prevalent. Crops containing many infected plants should be used for ware only. If a few plants show *Verticillium* wilt in a crop grown for " seed " they should be dug out and burnt, preferably with those on each side in the drill as otherwise a few infected tubers may be left and lifted with the crop. Also neighbouring plants may have become infected by root contact and if this occurs late in the season they may not show distinctive symptoms before the haulm begins to ripen off.

FUSARIUM WILT

It is associated with a brown discolouration of the vascular bundles throughout the plant, as in *Verticillium* wilt; indeed, the two diseases can only be distinguished with certainty by isolation of the fungus in pure culture. The *FusaHum* enters the tubers along the vascular bundles from the stolons and in the worst cases it grows through them all round the tuber so that a brown ring is visible when it is cut in half. During the winter the fungus slowly grows into the flesh and may cause rotting of the whole tuber. If those only slightly infected are planted they give rise to infected shoots which grow only into small stunted plants. *Fusarium oxysporum* requires a fairly high temperature; it makes scarcely any growth at 48° F. and its optimum lies between 77° and 86° F. In cool climates it is only likely to be troublesome in potatoes forced under glass. Warm, damp soils favour infection but drought hastens the death of the infected plant. Trouble is not likely to arise even in forced potatoes if they are raised from " seed " from a normal field crop. If the disease does occur potatoes should not be grown in the same soil for several years.

BLACK DOT

This very common fungus may damage potato haulm in dry seasons. The stems are attacked just below soil level, and become marked or even girdled by dark-brown or blackish lesions from which the skin breaks away. Girdling of the stems leads to yellowing and rolling of the leaves and premature dying of the haulm. The fungus is common on the tubers where it causes silvery grey patches covered with tiny black dots like those of silver scurf. It may also girdle the stolons and penetrate some distance up the stems along the vascular bundles. The tiny black sclerotia, sometimes bearing conidia on their surface, may be found on any of the diseased tissues. King Edward, Red King and Doon Star are considered to be susceptible varieties. The fungus is cosmopolitan and has been described under several other names, of which *Vermicularia variant* Ducom. is that most used on the Continent. It is regarded as a serious parasite in France but is of little consequence in cooler climates. The fungus is better known as a parasite of tomato and egg plant.

DRY ROT

During storage of potatoes numerous species of *Fusarium* may gain access to the flesh of damaged tubers, sometimes through wounds but mainly through tissues already killed by blight. Dead patches of tissue caused by sun scald or damaged by disinfectant or other chemicals may also afford entry to weak parasites. Susceptibility of the tubers to these rots increases during the winter. True dry rot, however, is caused mainly by one easily recognized species of *Fusarium* ^ *F. coeruleum,* so-called because its conidial pustules instead of being salmon pink have more or less of a blue tint. They may be covered with a white or pinkish spore powder but if this is rubbed off the blue colour will be found in the base of the pustules. It also sometimes appears in the mycelium, lining cavities in the rotting flesh. *F. coeruleum* is an active parasite though it may often gain access to the tuber through a scab wound or a bruise.The first sign of dry rot is a brown patch on the skin, which rapidly increases in size, becomes slightly sunken and surrounded by concentric wrinkles. The underlying flesh is deeply penetrated by a soft brown rot, full of cavities lined with white or blue mycelium, and giving off a peculiar sweetish smell. In time the rot destroys the whole tuber which becomes hard, dry, shrivelled and very light in weight. The characteristic conidial pustules burst through the dead skin.

SILVER SCURF

Sometimes silvery patches of skin are noticeable on potato tubers, especially when they are washed. The underlying tissue may be quite normal or a little darker than usual. Silver scurf is seldom seen at lifting time but develops in the clamp and may be regarded as a slight blemish rather than a disease. In wet clamps the condition may be widespread and extensively affected tubers may tend to shrivel if stored afterwards in a dry place.

Otherwise no real harm is done. Under warm, damp conditions a black velvety layer of short conidiophores and conidia appears on the silvery patches, springing from very tiny black specks in the skin, rather like the tiny sclerotia of *Colletotrichum atramen-tarium*. No control measures are normally called for, but if the fungus becomes undesirably prevalent a longer interval between potato crops should be tried, as infection of the tubers is derived from the soil.

PINK ROT

The fungus, which occurs in the soil, may attack the underground parts of the haulm and the stolons, sometimes causing the plants to wilt, but this usually happens only at the end of the season. Losses from the disease are due to infection of the tubers.

This occurs normally at the heel end while they are still attached to the parent plant, but the fungus may also sometimes gain access through an eye or a wound. Inside the tuber it causes rather a rapid rot but the diseased potato remains fairly firm until it is completely killed. When squeezed, however, sap generally oozes from it. If an infected tuber is cut and exposed to the air the flesh turns pink in a few minutes and within half an hour the whole diseased portion becomes salmon-coloured. Later it slowly turns purplish-brown or blackish-brown. Similar colouration may occasionally occur in healthy tubers or in those attacked by black-leg.

The infected tubers decay fairly quickly and are not likely to survive until planting time, hence there is little risk of spreading the disease with the " seed." As the causal organism is a soil-borne water-mould, thorough drainage of potato fields is desirable and a long rotation should help to prevent infection of the soil being built up to a dangerous level. All infected tubers should be destroyed, not dug or ploughed back into the land.

JELLY-END ROT

This is a physiological disorder associated with a form of second growth, especially in Arran Pilot. When second growth occurs evenly over the rose end of the tuber, so that it becomes abnormally elongated, the tissue at the heel end loses its starch and becomes soft, flabby and shrunken. Ultimately it usually shrivels up and falls off leaving an irregular scar.

WET ROT

Even at lifting, tubers may be found that are completely soft and rotten; as a rule they are from a black-leg plant or have been suffocated by flooding or puddling of the soil. Potatoes that have been attacked by blight and black-leg or have been damaged during lifting and bruised by rough handling in transit, packed tightly in railway vans or in a ship's hold, or stored under ill-ventilated conditions, may be transformed in a few days into a rotten mess. Numerous species of bacteria may play a part in this decay. If heating in the clamp is discovered in time and ventilation is supplied many of the tubers will merely develop black-heart. If not, and the temperature continues to rise, the potatoes will be suffocated and afterwards reduced to a soft, stinking, slimy mass by the common putrefactive bacteria that cannot attack living tissues.

VIRUS DISEASES

Virus diseases occur wherever potatoes are grown, though they only spread rapidly in climates favourable to their insect vectors, hence potatoes grown in some mountainous countries and in exposed wet maritime districts can be kept free from viruses without great difficulty. In commercial potato fields the virus-infected plants occur scattered throughout the crop because they arose from tubers which were already infected when they were planted. Secondary spread of the diseases within the crop is usually evident only late in the season or only in plants grown from the harvested tubers the following year. In lowland districts the decline in productivity resulting from virus infection would render potato production uneconomic if fresh " seed " tubers were not secured every few years from a suitable " seed " producing region. Tubers from a diseased plant mostly contain the virus or viruses that have attacked it but a few may sometimes escape infection. Once a tuber has become infected, it, and the plant grown from it, always remain diseased. Hence the primary source of potato virus disease is the planting of diseased tubers. It is impossible to tell from its appearance or by any chemical test whether a tuber is virus-infected or not.

The symptoms caused by infection by a potato virus vary greatly, partly according to the variety of potato infected, partly according to the time elapsed since infection (symptoms in the second years are often very different and more severe than those in the season of infection) and partly according to environmental conditions, for heavy manuring, especially with nitrogen, and warmth may " mask " the symptoms in the haulm. Carriers, varieties which can be or always are infected with a particular virus without showing any symptoms, are common among potato varieties. In practice potato virus diseases are usually referred to one of four categories, irrespective of the particular virus present.

These are leaf roll, mild mosaic, severe mosaic and aucuba mosaic. This is the most destructive of all potato virus diseases and the most difficult to control. Rolling of the leaves may also be caused by drought, waterlogging, fume damage, *Corticium solani,* black-leg, insect damage to the stem, eelworm and the like, but in all these cases the rolled leaves are limp or at least pliable, and usually it is the upper young leaves that are affected. In true leaf roll, derived from a diseased tuber, the lower leaves are rolled, point more or less upwards and are stiff and hard.

Mild mosaic is defined as a mild light and dark green mottle, in which the leaves are neither distorted nor reduced in size. It may be so faint as to be visible only on a cloudy day or when the leaf is held to the light. Often it is clearest in the young haulm, and less apparent in late summer and autumn. Mottling may also result from manganese or magnesium deficiency, but then the whole haulm is usually affected throughout a field or in considerable parts of it. Aucuba mosaic, so called from a fancied resemblance to the variegated leaf of *Aucuba japonica,* is recognized by clearly defined, bright yellow spots scattered over the lowest leaves. The leaves are not distorted and are of normal size, and the plant seems to suffer little from the disease. Somewhat similar variegation may result from spring frosts.

This includes all diseases of the mosaic type in which the surface of the leaves is roughened, crinkled, streaked with dark lines, distorted or reduced in size and in which the haulm is more or less stunted, especially in the year after infection.

OTHER VIRUS DISEASES OF THE HAULM

Paracrinkle is a peculiarly severe form of crinkle caused in the stock when a scion of King Edward potato is grafted on certain other varieties. All King Edward plants contain the para-crinkle virus without suffering any ill effect from it. Paracrinkle is purely a laboratory disease; it has never been seen in the field and no means of transmitting it other than grafting has yet been discovered.

Top necrosis begins by the appearance of numerous small brown dead spots on the youngest leaves. These rapidly spread and the shoot dies back from the tip. If any tubers were formed before the plant died, corky spots develop in them during storage and they either die before planting or, at best, give a weak shoot which soon dies. Top necrosis is very seldom seen in a field crop but it is of practical importance in two ways. In the first place different potato varieties react by top necrosis to different viruses. Hence grafting scions from a plant under examination to a selected number of indicator varieties is a standard method of identifying the viruses it contains. This method is particularly valuable in detecting "carried" viruses. Secondly, because the progeny of a plant that reacts by top necrosis are all killed, this is a self-eliminating disease. A virus cannot be perpetuated in a potato variety in which it causes top necrosis. Hence much attention has been paid to breeding new potato varieties which react by top necrosis to as many viruses as possible.

VIRUS DISEASE OF THE TUBERS

Net necrosis is the name given to a fine brown network in the flesh of the tuber. This is a reaction of certain varieties, notably Golden Wonder, in the year they become infected with leaf roll. That is, tubers with net necrosis produce plants with primary leaf roll but these do not, as a rule, yield more tubers showing net necrosis. A somewhat similar brown network may be the result of slight frosting of the tuber.

Tuber blotch refers to small brown spots or blotches in the flesh of the tuber, rather like those known as internal rust spot. The condition may be a result of infection with *Solanum* Viruses. Potato varieties all react in much the same way to *Solanum* Virus, the cause of leaf roll, though the symptoms are more pronounced in some, e.g. President, than in others like Great Scot. Their reaction to *Solanum* Virus 2 also differs in intensity rather than in kind. They react very differently, some by top necrosis, others by a mild mosaic or by showing no symptoms of disease at all.

RED PEPPER

LEAF SPOT

The spots are at first watersoaked but soon dry out and are then white with a dark-brown margin. When many spots occur on one leaf it turns yellow and drops. Spots may sometimes occur also on the leaf-stalks and fruit stalks. The fungus cannot penetrate the skin of the fruit but may grow into it through the stalk and then sets up a " stalk rot."

It is likely to appear sooner or later where the plant is only grown on a small scale as a glasshouse crop. Spots appear on the fruit, starting as small translucent or dark-red depressions, which quickly grow larger and gradually become lighter in colour. The pustules of the fungus form a dark coating on the diseased tissue and exude pinkish masses of conidia, which are spread by rain splashes or on clothes or tools of workers.

Sap Warts

When pepper is kept wet, especially if the atmosphere is nearly saturated or when great changes in humidity occur, it easily develops oedema. This takes the form of a large number of very small grey-brown warts on the underside of the leaf. One should try to avoid excessive humidity or much change in the moisture supply to the plant.

Grey Mould

We are not sure whether *Botrytis* is the primary cause of the damage or net. Quite likely the fruit suffered first from blossom-end rot and the grey mould followed and increased the damage. Leaves and stalks may also be attacked. The plants must not be kept too dry or the fruit will develop blossom-end rot, or too damp or *Botrytis* will be encouraged. Indeed the art of growing this, like other greenhouse crops, consists in supplying the plants with enough water in a damp enough atmosphere without letting them become moister than is necessary.

BLOSSOM-END ROT

Capsicum fruits may develop blossom-end rot like tomatoes. The first sign of the disease is a more or less sunken area at the tip, or close to it, on the half-grown fruit. The spot is usually a little paler than the surrounding healthy tissue but it may be dark-green and translucent. Gradually the spot grows larger but it does not cover the whole of the blossom half of the fruit as may happen in tomato. The spots may, however, become infected by fungi or bacteria, which set up a rot that may involve the whole fruit. The spots are not primarily due to a parasite, however, but to a physiological disturbance. Higgins has studied the problem closely in Georgia, both with controlled watering and under natural conditions out of doors.

In all cases it was shown that drought caused the appearance of the spots. In the open air they become infected by bacteria or fungi in the course of a few days. In fairly young spots neither fungi nor bacteria were found. Only as the spots increased in size did these organisms invade them and thrive on the already damaged tissue of the fruit. The spots start by collapse of the large, thin-walled cells round the ends of the vascular bundles. Red-pepper plants should not be exposed to drought, especially when the fruit is half-grown and has been growing vigourously.

BROAD BEAN

CHOCOLATE SPOT

Chocolate spot is one of the commonest diseases of broad beans and field beans. Dark brown spots appear on the leaves and stems. They may be small, more or less circular, and scattered, in which case little harm is done, or they may spread rapidly, cover the whole leaf surface and cause decay of the stems and death of the plants. In severe outbreaks very few pods mature. Brown spots occur on the pods as well and may penetrate to the ripening seed. Typical chocolate spot may be caused by the ubiquitous mould *Botrytis cinerea,* but there is also a specialized species, *B. fabae,* peculiar to broad beans, which causes a similar disease. Chocolate spot is favoured by moisture, ill-drained soil, shade and crowded planting. Hence it is a much more disastrous disease of field beans than of isolated rows of beans in gardens. It is also encouraged by potassium deficiency or by malnutrition of any kind.

MINOR LEAF SPOTS

Cercospora fabae Fautrey causes small chocolate-coloured spots or so across on broad-bean leaves. Under favourable conditions they become larger with a sunken grey centre and a raised dark-brown margin. In wet weather both sides of the spot may bear a silver-grey layer of conidia. Similar spots may be caused by the pycnidial fungus *Ascochyta fabae* Speg. This fungus may also cause light sunken spots with a broad dark-brown margin on the pods and is known to be seed-borne. Both *Cercospora fabae* and *Ascochyta fabae* are widespread but neither is usually considered important enough to warrant control measures.

SCLEROTINIA DISEASE

Broad beans, and especially field beans, are susceptible to a stem rot similar to that caused in other plants by *Sclerotinia sclerotiorum*. In the case of these beans, however, the cause of the trouble is usually a distinct fungus, 5. *trifoliorum* var. *fabae,* very closely related to the fungus of the well-known red clover rot. Although the race commonly found on broad beans has been distinguished from *S. trifoliorum* by small technical differences it can also cause rotting of red clover, and the typical *S. trifoliorum* can cause rotting of broad beans. Both fungi also attack common vetch, garden pea, sainfoin, lettuce, and possibly other plants.

BEANS—DWARF AND RUNNER

Anthracnose

Except in the warmest and driest parts of the earth dwarf and runner beans are liable to surfer severely from attack by the anthracnose fungus, the conidial stage of which is only too well known. It has become fairly common especially in cool wet summers, but the ascospore stage has only been found. All kinds of dwarf and runner beans may be attacked.

The disease is seed-borne and the seedlings may show brown stripes on their stems. In damp weather new spots readily arise, extending lengthwise into sunken brown stripes. On the leaves the spots generally start on the underside of the veins, though the surrounding leaf tissue may also be killed. The most conspicuous symptoms are those on the pods, which are circular or elliptical sunken brown spots with a darker edge. From these the fungus grows into the seed, in which it remains alive until germination, and then attacks the seedling.

On dark-coloured seeds it is difficult to see the spots resulting from infection. On white-seeded varieties the larger spots are pale brown but small ones may escape notice. The fungus overwinters in the soil, especially on dead pods and other plant fragments. Plants attacked in the seedling stage or soon after may be killed outright; at later stages the damage is mainly to the pods. The fungus fructifies as minute pinkish pustules of slimy spores on the surface of the spots and these are scattered, especially by splashes of rain and by handling when the pods are being picked. Enclosed localities where the air is apt to be moist and still favour the disease. Manuring with potash and superphosphate may make the plants a little less sensitive; sulphate of ammonia should only be used on a table crop, not in seed production.Bean varieties differ greatly in susceptibility so that considerable success has been attained in developing resistant types. Resistance may be dependent on one or more inherited factors but so far these have always been dominant in hybrids. It soon became evident, however, that the fungus, too, exists in numerous races different in infective capacity. Hence a bean variety may be resistant to several races of the fungus but not to others. Hence new varieties have to be exposed to infection by all known races of the anthracnose fungus before they can be described as resistant, and preferably this property should be combined with resistance to other diseases like halo blight, rust and mosaic. So successfully has this been done that some bean varieties now contain three hereditary factors for resistance and are resistant to 34 races of the anthracnose fungus. Actually they all become infected but in resistant varieties the fungus only penetrates the skin and then dies.

Fusarium Wilt

The chief symptom is a rolling inwards of the leaf edges, followed by wilting of part or the whole of the plant. The fungus occurs only in the vascular strands which are stained dark brown as in *Vertidllium* wilt, especially at the base of the plant. It is apparently not seed-borne and infection presumably comes from the soil, through wounds made by soil insects or by the mechanical splitting of the stem base by secondary roots. The optimum temperature for its growth is 79-82° F.

Foot Rot

Foot rot is caused by a soil-borne fungus prevalent in the Evesham district of Worcestershire. It attacks the roots and stem bases of dwarf beans, turning them reddish brown. The leaves turn yellow and only the first-formed pods swell properly. Foot rot causes most loss in dry seasons, when it may reduce the yield by one half. Strong-growing varieties like Flageolet are somewhat resistant as they put out fresh roots from the lower part of the stem to replace the diseased ones.

The same fungus has been found to cause a similar disease in French beans grown under glass. In this case sowing in sterilized soil or peat is recommended, followed by protecting the stem base with the rim of a bottomless pot when the seedlings are planted out in the house.

Rust

This is found on dwarf and runner beans all over the world. The cluster cup stage of this rust is white. This is usually followed by cinnamon-coloured pustules of uredospores on both sides of the leaf and finally by blackish-brown pustules of teleutospores. Bean rust is only occasionally reported. Sprays and dusts give no control of rust and if it breaks out the diseased plants should be burnt or composted and replaced by a resistant variety.

Red Nose

When beans are germinated in the seed-testing laboratory a slight pink discolouration is sometimes noticed at the micropyle or germ pore. This has been called " red nose " and is due to infection with the common fungus *Pleospora herbarum*. This fungus cannot cause disease of the growing crop but the infected seeds often decay or give sickly seedlings that die early. If they survive this stage they grow into normal healthy plants. Treatment with one of the proprietary organomercury seed dressings often gives improved germination.

Beans harvested under unfavourable conditions, or otherwise of poor quality, may be attacked during germination by bacteria and moulds like *Penicillium,* which reduce the percentage germination. Treatment with an organo-mercury seed dressing or with red copper oxide may give a definite improvement in germination.

Some beans germinate and develop their cotyledons (seed leaves) normally but the terminal bud or plumule is seen to be lacking. An axillary bud of one of the cotyledons may grow out to replace it but only gives a poor plant. The cause may be bacterial infection or damage at threshing, which has cracked the embryo just above the attachment of the cotyledons.

Microscopic examination will decide which cause is involved. Finally, beans may germinate badly (and cook badly) after a very hot, dry autumn or after storing in a dry warm place which hardens the skin of the seed. Storage in a damp warm place also brings about certain enzyme reactions within the bean and leads to formation of a " hard kernel," which also reduces germination and cooking quality.

Halo Blight

A number of bacterial diseases are known to affect dwarf and runner beans but their symptoms are usually somewhat similar so that they can only be distinguished by bacteriological investigations. The disease is usually attributed to *Pseudomonas phaseolicola.* This is seed-borne, and infected seed gives rise to stunted, spotted or yellow deformed plants which usually soon die. From them the bacteria are spread, probably largely by rain splashes, to healthy adjacent plants. On the leaves small, irregular, watersoaked spots appear, that are at first pale, but later turn brown with a transparent halo-like margin. Similar spots may appear on stems and petioles and from them small drops of bacterial slime may ooze out. When the pods appear they too become spotted and sometimes deformed, or even killed.

Seeds from infected pods are likely also to be infected with the bacteria. Sometimes they are spotted, wrinkled, or shrivelled, but they may bear no obvious mark of infection and yet give rise to an infected seedling.

It too is seed-borne and infected seedlings may be killed by it as they emerge above ground. On older plants it forms minute watersoaked spots or pale-green wilted areas which enlarge, dry out and turn brown. They usually have a yellow border. Reddish streaks appear on the stems, or reddish lesions may girdle the stem at the level of the cotyledons, or one of the lower nodes. The stem easily breaks at these points. On the pods there are dark-green, watersoaked spots that become dry, brick red and sunken.

LETTUCE

RING SPOT

This fungus causes oblong sunken spots, at first watersoaked, then yellowish and finally reddish or light brown, like small patches of slug damage, on the stem and the under-side of the midrib of lettuce leaves. Similar spots appear on the leaf blade, where they may be round or angular, defined by the larger veins. In later stages of the disease the tissue in the centre of these spots dies and falls out, so that the leaves look as though perforated by hail. The attack starts on the outer leaves and spreads inwards. Endive is also attacked by this fungus, so is the wild lettuce *Lactuca scariola* and probably the weed *Crepis.*

Bleaching powder is well mixed with ten times its weight of water and left to settle for half an hour. The clear liquid is then strained through muslin and the seed steeped in it for 4-6 hours, dried and sown at once. A small quantity of a wetting compound may be added to the solution. Other control measures include rotation of crops and destruction of the remains of all infected crops after the marketable heads have been cut. In greenhouses care should be taken to keep watering at a minimum and to avoid splashing soil on the plants.

Seedlings may be sprayed with a copper fungicide, but this is impracticable with maturing lettuce as the copper deposit on the leaves would render it unfit for human consumption.

OTHER LEAF SPOTS

The first symptoms are small light-brown spots on the leaves, usually round but sometimes angular or oblong. The spots enlarge, join up and may cause large parts of the leaf to become brown and wrinkled. On the outer leaves the centre of the spots may fall out so that they look like hail-damaged plants. Often the spots have a dark-brown edge or bear a number of darker concentric lines.

Septoria lactucae Pass, causes light-yellow irregular blotches, which turn brown and become covered with minute dot-like pycnidia. In later stages of the disease the spots may join together, or fall out as described for *Pleospora* above. The fungus is seed-borne and its pycnidia can be detected on infected seed by examination under a binocular microscope.

DOWNY MILDEW

Lettuce out of doors in frames and in greenhouses is often attacked by downy mildew. The upper surface of diseased leaves is marked by rather vague pale or brownish spots, the under-side of which is covered by a thin whitish mould. When conditions are favourable the fungus may spread very rapidly.

It has a low optimum tempe-rature, 43-50° F., and is hence favoured by cool weather and by low temperatures in houses and frames. Young plants are more susceptible than older ones. The fungus may overwinter in the soil but seems unable to survive severe frost. Morphologically similar downy mildews occur on many weeds of the lettuce family *(Compontae)* but it is unlikely that most of these can attack lettuce.

Two races of downy mildew on lettuce, one of which infects the wild lettuce *Lactuca virosa* while the other does not. Both occur on *L. serriola.* It is often impossible to establish where the infection of cultivated lettuce comes from, but one must remember that the conidia can be spread from a diseased crop in many ways, through the air, by adhering to boxes and baskets used for lettuce and on clothes or boots of workers. Seed-borne infection has not been demonstrated.

GREY MOULD AND RED LEG

Grey mould may cover a greater or smaller portion of a lettuce head and cause it to rot. This is especially likely to occur if the head has been damaged in some way and if there is a little dead tissue about, as from marginal spot or tip bum. In forced lettuce and outdoor white lettuce much more serious losses may arise by infection of the stem of young plants. This causes a reddish sunken wound, " Red-leg" which rapidly kills the plant, especially in overwintered seedlings transplanted in spring. Later in the season the stem wounds may heal and the plant mature more or less normally. Dusting with Folosan is said to afford some protection.

SCLEROTINIA DISEASE

These fungi gain access to withered leaves at the base of the plant and spread rapidly from them into the living head, which after a couple of days becomes limp and appears as though boiled. The heads may also appear to have fallen over, or to have been cut at the base. These fungi are easily recognized by their profuse development of white mycelium and their large, black sclerotia. The former species will attack almost any herbaceous crop, the latter has already been found on beans, asparagus, radish, etc. The clover rot fungus, common in red clover fields, also occasionally attacks lettuce.The control measures described for *S. sclerotiorum* are applicable to all these species.

RHIZOCTONIA DISEASE

Symptoms. When the heads are formed small rusty or chocolate-coloured spots appear on the under-side of the midrib of those leaves in contact with the soil. The midrib and leaf blade soon wither but the stem of the plant is less susceptible. Wilting of the outer leaves on sunny days is one of the first signs of the disease. Gradually the heads decay completely; at first they are brown and slimy, then they dry up and turn almost black. The brown mycelium of the fungus with its small dark irregular crust-like sclerotia permeates the dead tissues. The disease is favoured by damp and even more by high temperatures.

MOSAIC

The first symptom of lettuce mosaic is a clearing of the leaf veins, making them appear more prominent. Later a rather vague irregular mottling with shades of green may develop or the whole leaf may become yellow and hard-looking, as though damaged by wind or frost.

The leaf surface may be abnormally crumpled and its shape modified. There may also be tiny brown spots between the veins, or brown lines along the veins, and scorching of the leaf edges. The whole head is dwarfed and does not heart properly. When the crop " bolts " plants infected with mosaic remain stunted and may have brown spots on the stem and flower head. In cos lettuce browning of the veins is rare, but scorching of the leaf edges often occurs.

The virus is transmitted by lettuce seed and is spread in the field and glasshouses by aphides, including *Myzus persicae* and *Macros-iphum gei.* It can infect groundsel and sow thistle *(Sonchus asper)* and also sweet pea and culinary pea, causing a " streak " disease in the former and mottling of the leaves in the latter. It appears unable to infect dandelion.

TIP BURN

Shrivelling of lettuce leaf edges may be due to infection by *Pseudomonas marginalis* (Brown) Stapp and scorching to potassium or phosphoric acid deficiency. Probably more common than either is the physiological disorder known as tip burn. The disease begins as small brown spots near the edges of the leaves. As a rule they occur where the finer veins come together from the leaf edge. These small spots seem to hinder water from reaching that portion of the leaf which lies beyond them, hence it withers, turns yellow and dies so that a brown band appears along the edges of the leaves. Factors conducive to the production of dead leaf edges are: excessive moisture, great variations in the amount of moisture, lack of light, high temperatures and overmanured soil, especially an excess of nitrogen. Vigourously growing plants are more likely to develop the condition than those growing more slowly.

According to experiments, lettuce is sensitive to an excess of salt in the soil. Good results have been obtained by using fertilizers free from chlorine, viz. monoammonium phosphate and potassium nitrate. Experiments have also shown that varieties rich in sugars are less susceptible than those with a low sugar content. Steam sterilization of the soil predisposes to tip burn and that it is worst in heated houses and after prolonged dull weather. These factors are all conducive to soft growth, and tip burn sets in when the plants are exposed to strong sunlight

BEET

RUST

Beet rust is one of those which has all its stages on the same host plant. The orange-coloured cluster cups (aecidia) appear in May on the under-surface of leaves of young plants, and occasionally on the leaf stalks. This stage is seldom seen on seedlings but more often on the young leaves produced by overwintered roots kept for seed production. Spores from the cluster cups spread to young beetroot plants, on which successive generations of uredospores appear as rust-coloured pustules all through the summer. Towards the close of summer they become replaced by pustules of the dark-brown teleutospores.

Beet rust is most prevalent on mangolds and sugar-beet, but occurs also on garden beet, spinach beet and the wild sea beet *(Beta maritimd).* The most important control measure is to keep seed crops as far away as possible from fresh sowings. It may also be as well to inspect seed crops at regular and frequent intervals in spring as long as the cluster cups appear and remove the leaves that bear them. They should at once be put in a pail of lime water or other disinfectant fluid to prevent spores from being spread about the crop. Mother roots for seed crops should, if possible, be chosen from plants free from rust and stripped of leaf stalks which mighl bear the fungus.

LEAF SPOTS

Ramularia beticola Fautr. & Lamb, and *Cercospora beticola* Sacc. cause very similar circular leaf spots with a whitish centre and dark-purple margin on beetroot. On spinach beet the spots are pale with brown edges. When necessary, spraying or dusting with a copper fungicide should be practised. A leaf spot of mangolds caused by *Stemphylium botryosum* Wallr.

BLACK-LEG

The young seedlings turn brown at the soil level and shrivel up. Black-leg is worst on wet and sour soils which should be improved. Various fungi may play a part in the disease, including species of *Pythium* in the soil and *Phoma betae* Frank which is seed-borne. Hence treatment of the seed with the standard seed disinfectants may have a beneficial effect

BEET MOSAIC

Mosaic is very general in seed crops of mangold and sugar-beet, but seems less prevalent in garden beet, where it is only found in individual plants with conspicuously mottled leaves. In our infection experiments red beet proved not very susceptible and spinach beet even less so. Hence attacks are only to be feared when these crops are grown in the vicinity of sugar-beet or mangold seed crops. Beet mosaic will, however, attack spinach and hence overwintered spinach may be a source of infection from which the virus may be spread to beet by aphides. In our experiments the common goosefoot was not infected but wild sea beet did develop mosaic.

HEART ROT, DRY ROT AND CANKER

The innermost leaves turn black and die and gradually all the leaves on the plant may wither. Later new clusters of healthy leaves may appear on the crown outside the dead heart. On the root a slow, dry brown decay starts at the shoulder and gradually penetrates the tissues. In garden beet black spots may appear in the flesh.Heart rot and dry rot are much better known in sugar-beet and mangolds than in red beet. In the former they are known to be associated with limed soils and with dry years. Of late years excellent control has been obtained by manuring with borax at the rate of about 20 Ib. per acre. On soils where previous experience has shown that beet is liable to suffer from trouble of this kind borax should be applied in spring. Care must be taken not to exceed the recommended rate or the plants will be poisoned. Not more than i oz. to 12 square yards should be given. Watering with very dilute borax or boric acid solutions may check an outbreak that is just beginning.

BEET YELLOWS

Yellows is a common disease in sugar-beet and mangold crops where it affects all the plants in large patches or over whole fields. The virus may be spread by aphides to garden red beet, spinach beet and spinach and to goosefoot. Infected spinach beet develops stiff yellow leaves. Leaves of infected red beet are also stiff, but the yellow colouration tends to be masked by the red pigment in the leaves. Leaves of the diseased plants are gorged with starch, and are often prematurely attacked by the black mould *(Alternaria)* which normally only appears on the dying leaves in autumn.

The disease is not carried by the seed and there is no infection from the soil. The virus overwinters in beet plants, especially in seed crops, and is spread from them by aphides. As far as garden beet and spinach are concerned it seems to be prevalent only in the vicinity of sugar-beet crops.

CUCUMBER

ANTHRACNOSE

The seedlings may be attacked and fall over but this seldom happens. Leaf spots are the usual symptom at all stages of growth of the crop. They are pale green and transparent, later becoming dry, reddish brown, with a pale margin and tend to fall out. They enlarge more quickly than those of blotch and are more apt to fuse together into large withered areas, which may kill the whole leaf. Sunken dry patches, later becoming dry and floury, appear on leafstalks and stems. Sunken pale spots also appear on the fruit. The diseased tissue commonly bears pinkish coni-dial pustules of the fungus, easily distinguishable under a magnifying glass.

In greenhouses that have contained an infected crop the fungus occurs in the soil, on the glass, especially at joins, on decaying wood, old paper, etc.When the disease has occurred the houses should be thoroughly cleaned after the crop has been removed, sulphur should be burnt in them, all crop debris destroyed, the woodwork and glass should be sprayed with formalin 2:100 and the soil should either be renewed or thoroughly disinfected.

GUMMOSIS

When the seed has been sown in infected soil the seedlings may be attacked. As a rule the first symptom is a water-soaked spot on the hypocotyl below the seed leaves and often a dark stripe extending as far as the latter, on which dark spots may also develop. If the attack is severe the stem cracks at the attacked places. Under favourable conditions the infected seedlings may survive but the fungus is easily spread further from these lesions. In older plants oblong spots with yellowish-brown margins may develop on the stems and leaf stalks. On the leaves the fungus causes yellow or greyish-yellow spots which burst open. They are larger and more irregular than the small regular spots characteristic of blotch.

The disease on the fruit is the most general, most characteristic and best-known phase. Depressions occur in the skin, but several such spots often fuse to form large irregular patches. The spots often exude a gummy liquid, which hardens into a yellow or brown bead, entirely or partially filling the depression. This characteristic gummy discharge gives the disease its common name " gummosis". The conidia of the fungus form a grey or greenish-black layer over the depressions and on the edge of the spots.

The disease is caused by the mould fungus *Clados-poriurn cucumerinum.* Races of this fungus isolated from many different places varied greatly in their capacity for infection. It can over-winter in the soil, on woodwork, glass, etc.Cucumber seed extensively but was unable to demonstrate seedborne infection; its possibility is not, however, to be excluded. The fungus can grow at temperatures between 32° and 85° F. but thrives best about 70° F., whereas cucumber plants grow best about 80° F. If the atmospheric humidity is kept below 85 per cent development of the fungus is greatly checked. The fruits may also be attacked. There the spots are at first small, dark and somewhat sunken, not marked off from the surrounding healthy tissue by a sharply defined border. They quickly increase in size and may sometimes cover half the upper side of the fruit or more. Often the diseased tissues sink in to form a constriction in the fruit. This usually happens at one end of the cucumber; if it is at the stalk end naturally the entire development of the fruit stops.

MILDEW

Mildew may be very serious on outdoor cucumbers, it is of general occurrence but usually less serious on the hothouse crop. The disease begins as small white spots which may gradually coalesce to form a continuous powdery white coating over the shoot tips and leaves, particularly on the upper surfaces. When mildew is found in a greenhouse the ventilation should be improved and the old leaves removed as soon as possible. The remedies for mildew may also be tried usual.

BACTERIAL LEAF SPOT

This disease first appears as round or irregular transparent spots on the shoots and later on as well-defined, watersoaked, brown leaf spots, which dry up and fall out, just as in blotch. In the morning one often sees on the under surface of the spots small, clear, bacteria-filled droplets which dry to leave a chalky deposit. Watersoaked spots yielding a slimy white discharge may also arise on stems and leaf stalks. On the fruit the spots are at first very small, round and watersoaked, but later turn chalky white at the centre as the tissues die and dry out.

The disease is spread at pruning, by water splashes, and perhaps by insects. It can be transmitted on the seed, which may be disinfected by soaking for 5-10 minutes in corrosive sublimate solution, following by rinsing in clean water and drying. This disease is due to a bacterial infection of the vascular bundles and effects cucumbers, melons and gourds. The leaves develop pale green spots that quickly spread and are followed by a general withering of the leaves and shrivelling of the stems. When the latter are cut a sticky, whitish, bacterial slime exudes from the bundles.

This is a serious disease of outdoor cucumbers where it is spread by the knife at cutting and by insects, especially by a leaf beetle.

CHLORINE POISONING

Cucumbers have often been injured by being given stable manure, or latrine contents, to which chloride of lime had been added as a disinfectant. The leaves suddenly develop yellow areas at the tips, along the edges, or between the main veins. The newly set fruits suddenly stop growing and are spoilt; often they shrivel from the tip. Generally the injury appears at irregular intervals, i.e. it appears suddenly, then an improvement occurs, followed by another bad period, and so on. When one realises that the damage is due to injury to the roots by chlorine from the chloride of lime one can understand that the first outbreak follows application of the manure. Then when the chloride has been partly washed out the plants recover until damage to the roots again takes place.

One should make sure that chloride of lime is not added to the manure or latrine contents that are to be used. If one discovered that it has been added one should water the plants very heavily to try to wash it out as quickly as possible.

A tendency to bitterness in cucumbers is unfortunately quite common, especially amongst those grown in hothouses, but it has also occurred in cucumbers grown out of doors for sweet pickles. The bitterest fruits are dark outside and yellow inside but in most cases one cannot see from the outside whether a cucumber is bitter or not. The bitterness begins at the stem end and spreads through the flesh from there. There may be a noticeable difference in taste between two slices cut close to one another. A cucumber which is not bitter when pickled may become so later. Bitter cucumbers are most frequently met with from June to August. On the whole the smooth types are far more liable to bitterness than the spiny-skinned kinds. This is well known to many housewives and it is, therefore, easier to sell a spiny variety than a smooth one.Various factors encourage bitterness in cucumbers. There is some difference of opinion as to which is most important but it appears that sudden changes in growing conditions are of great importance. The phenomenon is often encountered when a period of sunshine follows a long period of cloudy weather, when the plants are too dry, or when they are very severely pruned. Other factors predisposing to bitterness are: damage to the roots by parasites, over-watering, and use of too strong manure or of fertilizers containing chlorine, wide variation in atmospheric or soil temperatures and, finally, setting of too many fruits. Individual plants may vary greatly in their tendency to bear bitter fruits.

One should endeavour to maintain steady growth by supplying heat during cold nights, ventilating and shading in strong sunlight, watering and cutting at the appropriate times, and avoiding too strong manuring. Bitter cucumbers are stained deeply by potassium iodide solution which may be useful in detecting them.

SHRIVELLED CUCUMBERS

Frequently cucumbers on hothouse plants cease growing and shrivel when about 2-4 inches long, without any obvious cause. The phenomenon is most frequent at high temperatures, with a moist atmosphere and wet soil. It also seems to be worst on plants carrying many large fruit. Several other conditions may stop growth of the young fruits: damage to the roots by chlorine, too strong farmyard manure, too cold water, drought, draughts, big ranges in temperature, or heavy pruning, leaving too few leaves so that transpiration is reduced.

STOPPAGE IN ROOT GROWTH

The roots may be damaged, so that they turn brown and no new roots are produced for a time, if too fresh manure or manure containing chlorine is used, or if the plants are watered with very cold water. This check to root development may cause injury to the fruits or of the leaves. A root rot of cucumbers, associated with species of *Fusarium,* the primary cause is believed to be bad cultural conditions, such as lack of drainage and poor aeration of the soil.

BLIND SHOOTS

Cucumbers in houses sometimes develop "blind shoots", i.e. the shoots suddenly stop growing and the tip ultimately withers, when side shoots develop to replace it. Even if these succeed, there has been a check to growth and, moreover, some side shoots may be destroyed also, so that it is some time before the plant but usually these are not the cause of blind shoots. The trouble is not due to parasites but to unsatisfactory conditions of growth which cannot always be precisely identified. It may occur where there is a draught from a hole in the glass, or from the ventilator. Good results have been obtained by putting cones over shoot tips exposed to draughts. The most frequent cause, however, seems to be that the plants were growing too vigourously, i.e. they had been given too strong manure, especially an excess of nitrogen and also too much warmth. Blind shoot tips are therefore most frequent on plants with large dark-green leaves.

The larger the plants are when planted out the better. Large plants have a shorter time in which " the top can go " than small ones that take longer to attain their full height. The plants should not receive extra nitrogen until they have begun to set fruit. Temperatures during growth should not exceed 86° F.

DROOPING LEAVES

Cucumber leaves may develop transparent withered edges and in the course of a few hours the condition may spread over the whole leaf, which collapses as though scalded. The stalk remains normal. In some cases the leaves may have been noted beforehand to be pale yellowish-green between the veins; in other cases there was nothing peculiar about them. This phenomenon is almost confined to the spring months. No parasite has been found associated with it, and after a time it ceases to occur, though affected plants suffer some check in growth. In one instance the damage seemed to be associated with the use of too fresh cow dung in the beds. Circumstantial evidence indicates that the trouble is due to some cultural fault. Cucumbers are a forced crop, grown at a high temperature and richly supplied with both food and moisture.

If, therefore, they are checked even for only a short period, by a sudden rise or fall in temperature, or if the roots are damaged by too warm or too cold water or by too fresh manure, they receive a shock which finds expression in some kind of malformation of growth.

SUDDEN WILTING

It frequently happens, generally in August, that outdoor cucumbers suddenly lose their turgidity and wilt. The trouble was particularly prevalent, when rain came after a long dry period. There was no sign of disease on the root stalk or leaves except that the plants drooped and eventually withered altogether..In a locality where the matter was examined more closely the phenomenon was equally prevalent in soil where cucumbers had been often grown and where there had been none for 10-15 years.

Attempts to correlate appearance of the disease with some cultural condition in the various districts were unsuccessful except that everywhere the plants were said to have withered " after rain". In some places they had been watered during dry periods and still many plants withered. Those that died were about to set fruit; they seem to be particularly susceptible to the trouble at that stage of growth.

Floral Abnormalities It happened at various places in Jylland that the outdoor cucumbers bore only male flowers or only an abnormally small number of female flowers. At one place where only Danish pickling cucumbers are grown, those raised in the greenhouse and planted out gave at first both male and female flowers and set normal fruit, but after a definite period the internodes became shorter, the leaves darker and only male flowers were produced. Some plants sown out of doors, which therefore developed later, produced only male flowers, while those sown in the hothouses and kept there remained normal throughout.

In another district the plants were of different varieties but the ovaries were not normal except on one or two and these set only a few fruits. Seed of the same varieties and consignments gave only normal plants in a third district.From a place where there were only a few female flowers, seeds were saved from the few fully developed fruits produced, and these were sown in the following year alongside seeds from normal plants. Seed from the abnormal plants gave almost exclusively abnormal plants while the others were normal. Thus the abnormality was inherited but its first appearance was due to some external condition. It may perhaps have been a sudden spell of sunshine after a dull period with low soil temperature that caused the appearance of male flowers.

MUSHROOM

BROWN BLOTCH OR BACTERIAL SPOT

Mushrooms grown in too warm and moist an atmosphere may be destroyed by purplish-brown spots on the cap, in which slimy drops may be found by closer examination. These drops teem with bacteria, which may be spread about the crop by various insects. Mushrooms -may be destroyed by this disease before they attain the size of a pea. The houses must be well ventilated after watering and the temperature should not rise above 55° F. Diseased fruit bodies should be removed and the attacked parts of the bed watered with chloride of lime.

BROWN PLASTER MOULD

Occasionally large crusts of felted mycelium appear on the surface of the manure and of the casing soil and slowly turn brown, owing to the production of minute sterile resting bodies by the fungus. These look like tiny yellow or brown sand grains. Brown plaster mould is introduced with the manure. If the manure is good and in good condition mushrooms will be produced in infected beds and even in the brown mould patches. On the whole, however, presence of the mould seems to reduce productivity of the bed.

GREEN PLASTER MOULD

On examining unfruitful mushroom beds one sometimes finds that the white mycelium of the mushroom has become surrounded by that of another fungus, which is at first also white, but soon turns yellow and ultimately grey-green. The fungus comes in with the manure and is said to be able to spread in sour wet manure. It may also be found on the surface of the beds. Direct control measures seem hopeless.

WHITE PLASTER MOULD

This fungus has a world-wide distribution and is of great importance in mushroom cultivation, especially where wet or too old manure is used. Then the fine white mycelium spreads in great quantity over the beds until their surface appears to have been powdered with chalk. Old patches of the fungus may take on a faint pinkish or greyish hue. Whether or not cropping of the mushrooms is affected by the plaster mould apparently depends on whether the latter succeeds in invading much of the bed before the mushroom spawn gets a start.

WHITE MOULD

This is a parasite of the fruit body in mushrooms and other gill fungi. Diseased fruit bodies are often transformed into unrecognizable lumps or, if the infection occurs comparatively late, they have a short thick stalk and hardly any cap. Their whole surface is covered by a close white mould and they soon break down into a stinking rotten mass. The term " bubbles " has also been used for this disease.

The fungus is said to be introduced with the casing soil, and use of fresh field soil from arable land is recommended. Local outbreaks in a bed can be treated by watering with formalin. The brown race of cultivated mushroom is resistant to white mould. The white mould layer of the fungus contains two-celled resting spores and sometimes also a conidial stage something like that of *Verticiilium.* The surface of the diseased caps was dark brown and the underlying flesh was stained yellow. *Dactylium dendroides* Fr. is a fungus attacking mushrooms and wild toadstools, which covers them with a close downy mycelium often with a pinkish tinge, and causes them to rot. It, too, probably comes from the casing soil.

VERTICIILIUM AND CEPHALOSPORIUM DISEASES

Several species belonging to these allied genera of moulds have been described as causing disease in cultivated mushrooms. Sometimes it causes shapeless, deformed mushrooms, covered with a fine white short mould, sometimes greyish-white spots on mushrooms of a normal shape, sometimes only white patches on the gills or stalk. This difference in symptoms depends on the stage of development at which the mushroom becomes infected. Even the most deformed fruit bodies do not rot but become firm and somewhat leathery in texture. Conidia of the *Verticiilium* are produced in vast numbers in minute slimy heads on the mould and are spread about the crop by *Sciara* flies and other insects. The fungus does not attack mushrooms at temperatures over 70° F. and seems unable to infect the white-capped strain.

It causes mushrooms to become deformed, spherical, or pear-shaped bodies like those resulting from mycogone attack, but dry, brown and elastic in texture.

Gill Mildew Disease

All the above described species of *Verticiilium* and *Cephalosporium* cause spots on the surface of mushrooms and, in favourable circumstances, reduce them to dry, misshapen lumps. The present fungus only attacks the gills, which tend to fuse together and have a mildewed appearance. Affected mushrooms are still edible but do not keep well, as their gills tend to rot.

Stable manure inevitably contains the spores of many kinds of fungi. Many of these are killed by the high temperature during composting, and if the mushroom spawn is introduced to the bed at the correct moment it usually grows away ahead of the remaining fungi. But under unfavourable conditions and if there are other sources of contamination, as from boots or tools, infection of the bed by competing fungi may forestall growth of the mushroom mycelium. Competition of this kind has already been mentioned under the heading of brown and white plaster moulds.

Between the mycelium of species of *Fusarium,* especially *F. oxysporum* Schlecht. and *F. solani* (Mart.) Sacc. var. *martii* (Appel and Wollenw.) Wollenw., and the mushroom mycelium there is, however, a reciprocal hostility or " antagonism." Growth of one mycelium excludes that of the other even at some distance. Presence of these *Fusaria* may lead to individual mushrooms becoming brown inside and tough, or to many small button mushrooms remaining undeveloped. Sometimes the mushroom growth may be restricted to a single crop and sometimes there is no crop at all. Sterilization of the casing soil by heat kills these moulds and hence prevents trouble from that source.

MUSHROOM BED INVADERS

The false truffle *Pseudobahamia microspora* Diehl and Lambert produces nodular, cream-coloured or pinkish hard fruit bodies, buried in the casing soil. *Xylaria vapouraria* Berk., called " seaweed " by the growers, forms tough, branched, black resting bodies in the casing soil. Its white mycelium permeates the underlying compost. The sclerotia send branches up to the surface of the casing soil, where they form pinkish bodies flush with the surface. The sclerotia are said to have a strong smell like cucumber, detectable by crawling over the surface of the beds and smelling the casing soil.

ROSE COMB (NON-PARASITIC)

After oil sprays have been used to destroy insect pests, or where oil stoves are in use, or woodwork has been coated with creosote, mushrooms sometimes bear a malformation on the cap somewhat like the rose-comb found in some breeds of poultry. The irregularly lobed excrescence is due to abnormal gill formation on the upper side of the cap as well as on the lower.

Prevention of Mushroom Diseases

Production of a healthy crop of mushrooms depends largely on the care taken in preparing and tending the beds.

The usual disinfectant is burning sulphur, but there are other methods whose relative effectiveness has not been fully ascertained. Empty houses may be sprayed throughout with formalin i: 50, preferably with a power sprayer, so that soil, walls and ceiling are thoroughly soaked. The house is then locked and kept warm for two days and nights, then thoroughly ventilated. The man who sprays should wear rubber gloves and his eyes and mouth should be protected. Boards and other woodwork may be disinfected by spraying or washing with copper sulphate or one can go over it with a blow lamp so as to burn off all loose splinters without scorching the wood. The copper sulphate treatment has an advantage over the blow lamp and formalin in that it leaves a small deposit of copper in the wood which preserves it without harming the mushrooms. Milk of lime is also sometimes used on woodwork.

If the compost beds are laid directly on the ground the latter may be covered with a layer of peat dust which may be watered with acetic acid at the rate of 2 gallons per 40 square yards. This acidified peat will prevent infection of the bed from the ground. Shelves used for mushroom growing may be washed in formalin and flamed with a blow lamp before the compost is placed on them. The quality of the manure is of the utmost importance even in disease prevention. The right consistency of the horse manure, not too fresh and not too old, fermented long enough but not too long, can only be learned by experience. The manure should be moist enough to hold together when squeezed into a ball but not so wet as to drip liquid when this is done. The moisture can be increased by watering, or reduced by mixing with ground gypsum.

The final fermentation (heating) should take place inside the mushroom house, and when it is at its height sulphur should be burnt to destroy fungi and insects on the surface. It is important to keep down mites and insects because, in addition to doing damage themselves, they spread fungus spores. The casing soil should be obtained from the subsoil. Surface soil contains weed seeds and often *Mycogone* and if it must be used it should be disinfected by formalin, or by steam sterilization. A sandy loam is the best type of soil and a neutral reaction is desirable. Steam sterilization sometimes reduces the yield of mushrooms, while after using formalin the house must be left to ventilate for some time.

The spawn is added to the compost as soon as the temperature 3 inches below the surface has fallen to 80° F., so that the mushroom mycelium may spread ahead of competitors.Diseased mushrooms should be removed with their mycelial strands and with the surrounding soil as well. Before removing them the spot should be watered with formalin i: 50, chloride of lime or raw carbolic acid. It is better to remove diseased fruit bodies wholesale than to see the whole crop destroyed by disease. The diseased material should be burnt. Gypsum dusted over the beds helps to prevent plaster mould. If watering becomes necessary it should be done through holes in the casing soil instead of by sprinkling the surface. The house should be kept damp by watering round about the beds. It is better to water between picking periods rather than during them.

If the soil has been very mouldy one may spray it with very dilute Bordeaux mixture. When greenhouses are cleared after a crop the earth or manure removed should not be put near mushroom houses or on the site where mushroom compost is to be made.

ONION

SMUT

By careful examination smut can be detected on quite young seedlings. It appears as opaque dark spots or streaks under the skin of the first and subsequent leaves and sometimes in the bulb scales of young onions. The leaves first attacked are often abnormally thickened and twisted. Sooner or later the skin over the streaks splits irregularly, exposing the black spore powder. These spores contaminate the soil and many remain alive there for twenty years or more. Many of the infected seedlings are killed outright by the disease; others survive and grow, but are malformed, while still others grow almost normally but develop fresh smut lesions on their leaves and bulbs. Individual plants may completely recover from the attack; diseased leaves and bulb scales are thrown off, and the bulbs ripen more or less normally, though they are always small.

The smut fungus which causes this disease can grow and infect seedlings at temperatures between 50° and 85° F. High temperatures after sowing therefore favour the fungus. Infection takes place from the soil and can occur only in the seedling stage, from about the second day after germination until the first leaf is developed. Older plants cannot be infected. The fungus enters that part of the seedling which is beneath the soil or just above its surface. It penetrates the epidermis and permeates the tissues as far as the growing point, after which the new leaves become infected as they develop. The incubation period is 5 days.

Wherever possible onions should be sown in soil free from smut spores. When they are big enough to be transplanted they can safely be transferred to infected land. It can attack many species of *Allium* and no resistant varieties of onion or leek are known. Various methods of dry seed treatment have been tried to avoid the troublesome formalin treatment. Good results were obtained with Brassicol but the method should be tested in larger-scale experiments before being generally adopted. The onion seed was slightly damped and then mixed with half its weight of Brassicol.

NECK ROT

It may spread in storage. The diseased scales become somewhat brown and transparent, looking as though cooked, and in a moist atmosphere they develop a short hairy coating of grey mould. Bacteria often follow as secondary invaders and set up a rapid soft rot. In the later stages of decay small black sclerotia form on the diseased part or even over the whole bulb. They may sometimes form a continuous scab, especially at the neck of the bulb. The disease is usually controlled when the bulbs are dried thoroughly and quickly after lifting and are then stored in a cool, dry, well-ventilated place. If they remain damp for long, neckrot may be very destructive. It is advisable to bend over the tops before growth has ceased. Onions are best stored at 35° F. Onions that have made very luxuriant growth seem more susceptible than those grown at a normal rate. White-scaled varieties are more susceptible than those with coloured scales.

DOWNY MILDEW

This disease does not usually break out until the bulb is about to form. Then a pale coating of the fungus appears on leaves and stalks. As the leaves wither the fungus turns brownish purple. In damp weather this coating may appear before there is any discolouration of the infected leaves, but they soon become pale and translucent and shrivel from the tip downwards. In mild attacks one may mistake this yellowing for normal ripening. Gradually the infected leaves wither to the base, often being replaced by new ones which are also usually attacked.

The dead parts are often covered later by a nearly black layer of the mould *Stemphylium botryosum*. In severe attacks the bulbs remain small, shrivelled or unripe. The disease also attacks shallots and in them it appears first on those grown from infected bulbs. The leaves are small and wither quickly, becoming pale, flaccid and covered by the greyish-purple layer of mildew. From these plants the disease spreads to the rest of the crop and causes symptoms like those described on onion. Both on onion and shallot the fungus grows down from the leaves into the bulb scales and persists there until the following year. Thus shallots and onion sets are a common source of infection. In severely infected bulbs the outer scales become brown, translucent and shrivelled. Some may die during the winter; others survive and sprout prematurely.Within the tissue of the infected leaves resting spores may be found, which can survive the winter and even remain alive for 5-6 years in the soil where onion tops have decayed. Cause and importance. Onion downy mildew is caused by the fungus *Peronospora destructor,* formerly often known as *P. schleideni* Unger. It can attack all varieties of onion and shallot, but is seldom seen on leek.

When onion sets are used it is important to obtain them from a completely healthy crop. If that cannot be done the sets should be graded and only the large sound ones planted and these should be carefully polished. The fungus in the sets may be killed by keeping them for more than 8 hours, or up to 24 hours, at a temperature of 104 F. In some experiments good results have been obtained by dipping sets in copper sulphate solution, or lime water, but there were no untreated controls for comparison. There is much to indicate that the fungus penetrates the bulb scales too deeply to be completely killed by such treatment. Seed should be harvested from sound plants.The soil should not have carried a diseased onion or shallot crop in the previous four years. Remains of a diseased onion crop should be burnt. When nearly all the onion tops in one onion-growing district were burnt for several years in succession, to control onion fly, the incidence of downy mildew was reduced at the same time. Onion beds should be well drained and exposed to sun and wind. Spraying and dusting are less effective than one might desire. It is difficult to get the liquid or powder to adhere to the onion leaves. If it is tried a spreading agent should be added.

WHITE ROT

The disease usually appears about the end of May or beginning of June, especially in autumn-sown crops. The older leaves turn yellow and fall over and the plants can be easily pulled up because their roots are rotten. Round the base of the bulb there is a soft, flurry, white mycelium, hence the name " mouldy nose" sometimes applied to the disease. In later stages this mould becomes matted and innumerable tiny, round, black sclerotia, none bigger than a pin's head, develop in it.

They are readily detached and left in the soil where they may remain alive for many years. All varieties of onion may be attacked, though some are more resistant than others. The fungus also attacks shallot, leek, and garlic and may occur on the wild crow garlic so that land may be infected with it that has never carried an onion crop. The sclerotia may also be found in seed samples and spread in that way. Probably the most important means of spread, however, is by infected sets and transplants. After an outbreak of white rot, onions, shallots and leeks should not be grown on the same ground for at least 4 years, preferably not for 8 or 10 years. Infected plants should be burnt.

SHANKING

This disease of onion and shallot closely resembles shanking of tulip. The outward symptoms are yellowing and shrivelling of the leaves and softening of the base of the plant. Some or all of the roots are soft and watersoaked. When-an affected bulb is cut longitudinally the bases of the youngest leaves and the tissues of the growing point are found to be discoloured, soft and often contracted. The light-green or white healthy tissue contrasts sharply with the dull buff infected area.

In some cases infection has been shown to occur from the soil in which the seedlings were raised prior to transplanting. This is not always the case, however, as the disease may occur hi onions drilled out of doors. Onion seed should not be sown in soil that has carried a tomato crop, as the *Phytophthora* spp. which cause foot rot and damping off of tomatoes are the same as cause shanking. Onions for transplanting should be raised in sterilized soil.

ROOT ROT AND FOOT ROT

Various species of *Fusarium* may attack onion, both seedlings and older plants. Roots of seedlings often become attacked by *Fusarium* from infected seeds, and they turn yellowish-pink and decay. The plants put out new roots which are often also attacked. The decay spreads upwards from the roots, and if the bulbs have formed the scales may rot at the base and the leaves turn yellow at the tip. The decayed parts of the bulb become reddish-brown or grey-brown, dry, mealy and overgrown by a pink or white mould; finally they become dry, shrivelled mummies. Alternatively there may be secondary infection by bacteria, causing a-soft, wet, stinking rot. Slightly infected bulbs may be overlooked at lifting and decay in storage.

MOULD ROT IN STORAGE

During storage onions may be seen with soft or dry areas on which there is a sparse blue-green covering of mould, usually beneath the outer scales. This is a species *of Penicillium. A* similar trouble occurs on garlic. The powdery black mould *Aspergillus niger* van Tiegh may occur in a similar situation and in Southern Europe a mould with yellowish-brown spore clusters causes rotting of garlic and onion bulbs. These moulds attack bruised bulbs or those damaged by frost or sunscald and stored in ill-ventilated places.

VIRUS DISEASES

These are common in shallot, much less so in onion, and are commonly referred to under the names yellow dwarf and mosaic, but the viruses causing them have not been extensively studied. Possibly some forms of the disease, however, may be caused by the common Cucumber Mosaic Virus, which has been shown to be transmitted by the shallot aphis, *Myzus ascalonicus* Doncaster. Infected shallots are small with twisted, compressed, limp leaves, usually streaked with yellow. They make only small bulbs, which may be soft and not keep well. This condition is general where half a crop may be affected. It is at least as common in English shallots and occurs to some extent in onions, especially in seed crops or in the vicinity of diseased shallots. Similar symptoms have been seen in the perennial Welsh onion, *Allium cepa* L. clon *perutile* Stearn. The shallot aphis occurs on stored bulbs as well as in the field, so possibly some transmission of the disease occurs in storage. Cucumber Mosaic Virus is also transmissible mechanically, as by bruising sets or transplants. Soil infection and seed-borne, infection have not been proved. Infections may not give rise to visible symptoms in the first year but the bulbs will yield diseased plants if replanted.

Onion and shallot sets should only be taken from healthy crops. If the health of a consignment is suspected a sample may be planted in a greenhouse during the winter, if infected bulbs are present they will give obviously diseased foliage. It was apparently this disease that Buchanan cured by keeping the bulbs at 84° F, for 112 hours.

TURNIP AND SWEDE 16

Brown Heart

This disease, known as "Raan" in parts Brown heart tends to be worst and by a name meaning roots. It is caused by a deficiency of available "marbling "affects only the boron in the soil. In some districts, notably flesh of the root. The soils are often halfway between the centre and the rind is truly deficient in boron, but in other districts seen to be marked with irregular discoloured brown heart may be the result of temporary boron deficiency brought about by excessive liming.

In such cases use of an acid fertilizer like sulphate of ammonia in place of sodium or potassium nitrate may have a beneficial effect. Brown heart can be easily and cheaply prevented by supplying boron to the soil, preferably before the seed is sown, in the form of borax or boric acid. Because boron is poisonous to plants when present to excess the dressing should not exceed 20 Ib. borax or 15 Ib. boric-acid per acre, or i oz. to about 15 and 20 sq. yards respectively, and must be thoroughly and evenly mixed with the soil. It may also be applied mixed with the fertilizer, though preferably not with sulphate of ammonia.

PEA

LEAF SPOT, POD SPOT, AND FOOT ROT

The common leaf and pod spot fungus *Ascochyta pisi* causes small yellow brown spots with brown edges, dotted with small brown pycnidia, on leaves, stalks and pods. The attack is general and may be destructive. Very similar fungi occur on sweet pea and on broad bean but it are not known whether these can infect peas.

If the fungus attacks the fruit stalk the pod falls. From the infected pod the fungus often grows into the seeds. Hence the disease is commonly seed-borne. Infected seed may rot or may give rise to a seedling bearing diseased spots on the stem and the tip of the young shoot. These may die before emerging above ground or may survive and suffer from root rot, depending largely on the soil temperature and speed of germination.

Two closely related fungi, *Mycosphaerella pinodes* Vestergr, which has an imperfect stage *Ascochyta pinodes* Jones, and *A. pinodella* L. K. Jones, also cause spotting of pea leaves and pods and are important causes of foot rot. The leaf spots of *M. pinodes* are small, dark purplish brown, often with concentric rings.

On the stems it causes purplish streaks and on the pods small purplish spots which may combine to form large blotches. *A. pinodella* causes small spots like *M. pinodes* but less numerous.

ROOT ROT AND FUSARIUM WILT

Root rot may sometimes be due to seed-borne fungi, but usually a poor seed bed and too frequent cropping with peas are important contributory factors. Repeated cropping with peas may succeed for a few years, but the only safe method of cropping is a long rotation to prevent the build-up of soil infection. Once soil has become heavily contaminated with root-rotting fungi it is difficult to get rid of them and several of the fungi concerned are not particular about the plants they attack. The best plan is to retain the name root rot for a disease in which either the roots or foot of the plant are attacked by a decay starting from the outside and turn brown or black.

Fusarium wilt *(Fusarium oxysporum* Schlecht.) is a distinct disease, in which there is no discolouration of the cortical tissues in roots or stem base but the vascular strand turns brown, the plants become yellow, either individually or in patches, and in severe infections they wilt and die. Affected plants yield very little. It is encouraged by heat and is therefore worst in warm climates. Root rot and *Fusarium* wilt not only cause loss of crop but lead to difficulties in canning peas, because the peas ripen irregularly and also tend to be abnormal in composition and hard when canned.

Seed disinfection may be helpful in certain cases, but the chief emphasis is to be laid on maintaining healthy, fertile soil, including the practice of a long rotation of crops. In Wisconsin addition of a complete artificial fertilizer to the seed drill has led to quicker germination and less root rot. Excess nitrogen and phosphoric acid makes peas more susceptible, while excess potassium makes them fairly resistant to *Ascochyta* and *Mycosphaerella*. Ensiling pea straw is insufficient to destroy root-rotting fungi on it; the straw from a diseased crop should be burnt.

DOWNY MILDEW

On the leaves of field and garden peas this fungus causes yellowish blotches covered on their lower surface by a whitish or pale-mauve layer of conidiophores and conidia. The infected areas may eventually turn brown and in some cases the whole plant may be permeated by the fungus. The pods may be attacked and bear pale-green or light-brown spots. These seldom produce conidia but the resting bodies, oospores, of the fungus are formed within the tissues of the pod. Infected pods often become spotted and deformed but it is not possible to tell whether infection of the seed has taken place. The fungus overwinters in the soil and can infect pea seedlings as long as the soil remains cool, with a temperature below about 65° F.

Spraying with Bordeaux mixture plus a spreader has been suggested as a means of control but must be carried out in the early stages of the disease if it is to be effective. Rotation of crops will help to keep down infection from the soil.

MILDEW

The race of this fungus found on peas does not attack other plants and hence Continental authors often treat it as a distinct species *E. Pisi* DC. It causes the usual white powdery mildew on both sides of pea leaves, formed by the superficial mycelium of the fungus and its chains of conidia. The tiny brown perithecia are much less often seen. In early and severe infections the mildew may occur also on the pods. As a rule pea mildew, though conspicuous, causes little damage. In early infections dusting with sulphur or spraying with lime-sulphur may be tried. Severely infected straw should be burnt or ensiled. Resistance to mildew exists, based on several inherited factors, so that resistant varieties could no doubt be bred.

MARSH SPOT (MANGANESE DEFICIENCY)

In samples of pea seed, especially in that grown on wet soils, one often finds a more or less circular brown depression in the centre of the inner face of each cotyledon. This occurs especially in varieties with large wrinkled seed but there is no sign of disease on the seedcoat, or until the pea is cut across. Often, too, the tip of the tiny shoot, the plumule, is damaged, or when an affected pea germinates the tip of the shoot may die when only a few inches high. Usually secondary lateral shoots develop to replace the dead tip, and if the seed is sown in fertile soil the plant ultimately suffers little from marsh spot. The trouble has been shown to be due to a deficiency of manganese and may be worst when ripening is slow. It may be prevented by applying manganese sulphate to the seed crop at the rate of about i to *I* cwt. per acre. Spraying the foliage with manganese sulphate solution during flowering may be even more effective.

PEA POOR GERMINATION

Poor germination of peas may be due to pre-emergence blight caused by *Ascochyta,* or to rotting of the seed by various soil fungi, such as species of *Fusarium, Botrytis, Pythium* and or with one of the proprietary organo-mercury various bacteria, especially if the seed was of seed disinfectants. Protection against these rotting germinates badly when sown too early.

RHUBARB

RUST

The cluster-cup *(aecidium)* stage of this rust occurs on rhubarb leaves. It causes carmine-red spots on the under-side of which is the tiny whitish cup with delicately fringed edges. The same stage occurs on leaves of sorrel and dock. It is common in the vicinity of swamps and ditches but not elsewhere. The uredospore and teleutospore stages of this rust occur in summer and winter on the common reed *(Phragmites communis)* where they cause yellowish-brown and black oblong pustules respectively.

DOWNY MILDEW

It has been found to be widespread but does little damage and is therefore seldom noticed by rhubarb growers.Discoloured spots appear on the leaves and eventually turn pale with a red margin. Their size varies greatly. The usual layer of downy mildew conidiophores appears on the lower surface. On young spots they may form a continuous mauve coating but on old spots they occur only as scattered white specks. If the disease seems to be spreading the leaves should be sprayed or dusted with a copper fungicide. As a general rule there seems no necessity for any control measures to be undertaken.

LEAF SPOT

This fungus causes circular brownish sharply denned spots on the leaves, with a fine whitish mould on their under surface. If it appears to be on the increase a fungicidal spray or dust may be applied. The disease is only occasionally reported in Britain.

CROWN ROT

It fairly often happens that rhubarb rots at the crown, either starting in the crown and working a little way up the leaf stalks starting in the crown and spreading down into the roots. The leaves of such plants turn red or wither completely. Often the diseased tissues are covered with a layer of grey mould but it does not always follow that grey mould was the primary cause of the trouble. Other fungi or bacteria may have been the primary parasites which opened the way for the *Botrytis.*

Verticillium albo-atrum may also attack this crop, causing the plants gradually to wither and die. The trouble is usually ascribed to *Bacterium rhaponticum* Millard, which causes the roots to become soft and almost chocolate-coloured. The problem is complicated by the activities of the eelworm *Anguillulina dipsaci* Kiihn which is commonly associated with *Bacterium rhaponticum* in the rotting crowns and is apparently able to cause decay of the tissues on its own account. When a new rhubarb bed is made the mother plants should be very carefully selected and no part of a plant with any sign of crown rot should be used for planting. Also rhubarb should be replanted straight away after being dug up, not, as is so often done, left for some time on top of the soil.

CARROT

VIOLET ROOT ROT

The fungus has a reddish-purple mycelium which grows in a network of fine threads over the underground parts of diseased plants, which gradually collapse as the tissue beneath the mycelium decays. It may grow through the soil to the next plant and attack it also. In dry weedy soil the fungus may persist in the same place for many years. Many different kinds of plants may be attacked. Potato, turnip, swede, sugar-beet, mangold, broad bean, red, white and Alsike clover, asparagus, carrot, celery, chicory, parsley, salsify, seakale, carnation, holly, and the weeds wild carrot, plantain and creeping buttercup. It is the most injurious root disease of sugar-beet and becomes a serious problem where carrot growing is practised intensively.

Baudys and others consider that different races of the fungus exist and call that on carrot/. *dauci.* This view does not find general acceptance and it is wiser to consider the possibility of infection spreading from any of the above-mentioned crops to any other. The fungus persists in the soil for many years and little can be done to get rid of it once it has appeared in a field. Long rotations, avoiding root crops for a considerable number of years, offer the best hope of control. On a small scale it may possibly be killed by soaking the soil with corrosive sublimate solution at the rate of 2 oz. sublimate plus 18 oz. saltpetre to 22 gallons of water.

CARROT FUNGUS

The disease appears in late summer on one-year-old carrots as sunken grey spots on the upper part of the root, often forming a sunken ring round the rosette of leaves. Many blackish pimple-like pycnidia appear on the surface of the spots, from which there ooze out in damp weather red globules or tendrils, composed of innumerable conidia embedded in mucilage not sink in as much as in Rostrup's description. The spots became blackish-brown, covered with a short loose grey mycelium mingled with white mycelial tufts. Pycnidia were gradually formed in the grey patches. The fungus may spread freely during storage and may grow from infected roots into healthy ones and destroy a whole consignment. Numerous other fungi *(Stemphylium radicinum, Sclerotinia sderotiorumy Botrytis cinerea, Fusarium* spp.), however, often cooperate in this decay.

The fungus does most damage to the seed crop. When infected roots are planted out to seed, spots gradually appear on the stem and leaf stalks, especially at the axils of the leaves. Often the spots are situated one above another and run together to form a streak down the stem because the mycelium tends to spread along the furrows. Infection becomes widespread by means of the conidia during the summer. When flowering commences the disease has usually advanced so far that the root decays, the stem withers, the leaves hang down and the plant dies without setting seed. Seeds that do mature may also be infected and bear small black pycnidia.

The disease was formerly much more important than at present. Commercial seed production is now usually from carrots that overwinter where they were sown and are less exposed to the disease than carrots taken up in autumn, stored over winter and replanted in spring.

BLACK ROT

Dry blackish spots appear on the leaves, leafstalk and stems of growing plants but the phase of the disease most often seen is a slow rather hard blackish rot on the stored root. It usually occurs on the neck but may also occur farther down the side of the carrot. Under dry conditions the rot is slow and dry but under damp conditions it becomes a wet rot covered with a blackish-grey mould *Fusarium* sp. and other secondary fungi often appear on the decaying spots. In carrots planted for seed the tops may wither. The disease is seed-borne and kills seedlings by rotting the roots and hypocotyledonary stem.

DOWNY MILDEW

The upper surface of the leaf bears yellowish spots which later turn black and may cause the leaves to shrivel and fall. On the under-side of the spots is a loose white mould. This fungus is much more common on parsnip and races of it exist on many wild umbelliferous plants such as goutweed, hedge parsley and angelica. These should not be tolerated in and around gardens as, even if the race of mildew attacking them does not infect carrot, they may also harbour the carrot fly. An attack of downy mildew may be checked at the outset by spraying or dusting with a copper fungicide. Remains of infected plants should be burnt or otherwise destroyed in autumn.

CRACKED ROOTS

Boron deficiency has not been reported in English carrot crops. In pot experiments with carrots on moorland peat, however, 0delien in Norway found the roots very much cracked. Manuring with small quantities of boron compounds prevented the cracking. On slightly acid peat soils in Canada longitudinal cracking of carrots has similarly been controlled by manuring with 20 Ib. borax per acre. Carrots grown in sand cultures without boron also developed wide cracks, and leaves which turned yellow round the edges. Longitudinal cracking may also be due, however, to a sudden acceleration of growth when rain follows a dry spell.

ROOT ROT OF CELERY AND SCAB OF CELERIAC

These diseases are of general occurrence but seldom attain any real importance. In celeriac, brown smooth spots appear on the skin of the crowns and gradually develop a cracked, warted skin. In early and severe attacks development of the crowns is checked and their upper surface becomes covered with a brown or yellowish-brown, scabby skin which may crack, and the crowns do not keep well. This condition, too, is sometimes called "rust" but it has no connection with true rust caused by *Puccinia apii.* In the scabby patches tiny dark pycnidia are formed and these may also be found on seedlings at the base of the leaf stalk and occasionally on the seeds and leaves. In severe attacks seedlings still in the seed box may develop damping off symptoms, i.e. they rot at the soil level and die. Celery is attacked by *P. apiicola* mainly at the base of the leaf stalks and the rot spreads thence into the crown and the roots. The decayed leaf stalks are bluish-green or black and the rotting roots are brown.

Celery roots may also sometimes be attacked by soil fungi such as species of *Pythium,* causing a reddish root rot, *Thielaviopsis basicola,* causing a black root rot, and the ubiquitous *Corticium solani.*

HEART ROT AND OTHER BACTERIAL DISEASES

Bacterial soft rot involving the heart leaves and spreading up the leaf stalks and down into the stock is not uncommon in celery. The cause is probably the common soft-rotting bacterium. Damage by the fly larvae may begin the trouble and it is likely to become prevalent when there is no proper rotation of crops. Danish experiments indicate that excess lime and boron deficiency are another important cause of heart rot. A non-parasitic form of black heart is widespread in celery fields. It seems to follow periods of either drought or excessive soil moisture, especially if the crop is growing rapidly. A purely bacterial disease is that caused by *Pseudomonas jaggeri* (Jagger) Stapp. This organism causes small rusty-brown spots on the leaves. It can be controlled by dusting the seed beds weekly with copper lime dust and in the field crop by spraying with Bordeaux mixture.

MOSAIC AND OTHER VIRUS DISEASES

Bitterness

Celery and celeriac belong to the group of plants which profit by chlorine. Hence plentiful manuring with a chlorine-containing potassium fertilizer should give a good flavour and good keeping quality (high sugar percentage, low nitrogen content); the bitter taste comes from too high a nitrogen content.

In celeriac boron deficiency results in hollow crowns with dark tissue lining the cavities. The crowns tend to turn black when cooked. In severe outbreaks there is only a thin layer of normal tissue surrounding a large hollow.

In some cases there is also a yellowing of the leaves, particularly of the heart leaves, which may die completely and the tops have a tendency to lie along the ground. In such plants there is usually a cavity in the heart extending down into the root stock, into which one can stick one's finger. In other cases there is no outward sign of the damage, which is only discovered when the crowns are cut.

The disease appears to be purely physiological in nature and to be connected with particular soil and weather conditions. It seems most serious where a good growing period has been followed by drought and then by a further good growing period. Different strains of celeriac differ in susceptibility to heart rot. A high soil reaction clearly favours the trouble and it can be prevented by manuring with borax. At the experimental station at Hornum about i oz. of borax to 36 square yards entirely prevented the appearance of these hollow crowns on ground where untreated celeriac was severely affected.

This kind of hollowness seems to have no connection with the small hollow so often found in celeriac as to be regarded as more or less normal. This latter is often no larger than a hazel nut or a walnut and the surrounding tissue is quite white, though it may be a little soft around the cavity. The size of the cavity may vary somewhat with variety and conditions of growth but is much less than that due to boron deficiency with its unsightly dark walls.With ordinary celery in Florida, boron deficiency has been found to cause browning of the heart leaves and the appearing of brown cracks and lines all over the ribs of the outer leaf stalks.

Rusty Markings in Celeriac

The swollen crowns of celeriac are sometimes stained by reddish markings either in winding lines like worm tracks or in flame-like shapes. This red colouration is often, mistakenly, called " rust," but it has nothing to do with the rust fungus. It is therefore best to restrict the name celery rust to the disease caused by *Puccinia apii.* Unsatisfactory manuring and sour soil may co-operate to produce these golden-reddish patches in the crowns. Very close planting has the same effect. Apparently also other but in recent experiments no clear differences factors, as yet undetermined, have some in- have been demonstrated in the behaviour of fluence on it. celeriac grown with different manurial treatments.

CHERVIL

Foot Rot

Under unfavourable circumstances young chervil plants may develop a constriction at the soil level due to infection by one of the damping off fungi, fall over and die.The lower surface of chervil leaves may become quite white with a layer of coni-diophores of this fungus. The corresponding parts of the upper surface show as yellow spots which later turn black and may cause the leaves to wither. Races of this fungus attack many cultivated and wild umbelliferous plants. Weeds belonging to this family should, therefore, be destroyed in and around the garden. In the early stages of an outbreak an attempt may be made to stop it by spraying with a copper fungicide. The remains of an infected crop should be burnt or otherwise destroyed in autumn.

SPINACH

ANTHRACNOSE

This fungus causes olive-green angular leaf spots that may turn yellow or almost white and dry up. They may join together especially along the leaf edges and cause extensive withering. *Colletotrichum spinaciae* seems to be as yet unknown but was found on spinach. It is not uncommon on spinach seed and may be very injurious to germination. Seed-borne infection up to 60 per cent of a sample has been found. Fungicidal treatment by shaking up the seed with an organo-mercury seed disinfectant should be tried. It may also be useful to spray the seed crop with Bordeaux mixture or similar fungicide.

LEAF SPOT

The early stages of attack consist of small, rounded, white spots, scattered over spinach leaves, mainly over the apical half. In later stages the spots are more numerous and may join up and are more or less covered with a fine olive-green or blackish mould. The first spots are usually on the upper surface of the leaf but in later phases of the disease the fungus is equally abundant on both surfaces. No control measures have been suggested.

DOWNY MILDEW

Irregular, ill-defined yellowish patches appear on the upper surface of the leaves, while on the corresponding parts of the under-surface there is a close, downy, grey mould that may acquire a lilac hue. Severely infected leaves are unsaleable.

The fungus may overwinter as oospores in the soil and these have also been found mixed with the seed. A high degree of humidity, at least 85 per cent, is needed for production of conidia and these germinate only in the presence of surface moisture. Infection may occur at temperatures between 40° and 75° F. with an optimum at about 48° F. At temperatures of 60-65° F. infection may take place within 3 hours of inoculation.

Spinach should not be grown on land that has already carried a diseased spinach crop and it is advisable to sow the spring crop at some distance from overwintered spinach. With indoor crops the temperature should be kept fairly high. On seedlings one may try spraying or dusting with a copper fungicide, taking care to cover both sides of the leaf. This is impracticable with older plants whose leaves are ready for table.

VIRUS DISEASES

Spinach Blight

This common virus causes a very destructive disease of spinach. Young leaves become yellowish, thickened, brittle and remain small with the edges inrolled and wrinkled. In severe cases some leaves may be reduced to a fine ribbon and others twisted, distorted and mottled. The outer leaves later turn yellow, then brown and shrivel until only a central cluster of yellow malformed leaves is left. When seedlings are attacked they become completely yellow or die.

The virus may be transmitted mechanically with the sap or, as occurs in nature, by aphides.

Mosaic

Mosaic of spinach caused by infection with the common beet mosaic virus from beetroot or sugar-beet is well known and has been produced experimentally. The first sign of infection is a downward arching of the young leaves, accompanied by the appearance of numerous bright golden-yellow spots. The yellow colour gradually spreads, and then the outer leaves die back beginning at the tip. The virus is spread by the beet aphis and other species and can be transmitted artificially by sap inoculation. The incubation period is 1-3 weeks.

White goosefoot or fat hens *(Chenopodiwn album)* may also be infected by this virus. It develops isolated yellow spots and the leaf edges curl under. The sow thistle *(Sonchus arvewis)* may also become infected. These weed hosts may play a part in enabling the virus to survive between beet and spinach crops.

STRAWBERRY

MILDEW

Symptoms: Mildew may attack strawberries out of doors or under glass. When the leaves are attacked they bear only a thin coating of the floury white spores among the hairs on their under surface. On the corresponding parts of the upper surface dark purplish blotches appear and the leaves roll upwards, exposing their under surface. On the fruit mildew produces a similar sparse white powdery coating and if they are infected before they ripen them do not turn a good red colour but become dry and poorly coloured. Their market value is thus impaired.

Wide variations in temperature appear to encourage the disease and should be avoided under glass and in frames. The usual sulphur fungicides may be applied until the fruit begins to ripen and after it has been picked.

LEAF SPOT

Reddish-purple spots appear on the leaves which become more or less circular and turn white at the centre, which may die and fall out. In severe infections the spots may be small and angular, but crowded together so that large parts of the leaves turn brown and shrivel. During the summer a white mould composed of bunches of short hypha: bearing conidia of the *Ramularia* stage appears on the centre of the spots. On overwintered leaves the tiny black *Mycosphaerella* perithecia may develop, sunken in the leaf tissue. They are, however, seldom seen.

LEAF BLOTCH

This disease starts as small purple spots like those of leaf spot but never develops the white centre so characteristic of that disease. The spots gradually increase in size up to an inch or more and may involve an entire leaflet. The large spots are dark-brown, often paler at the centre, with a dark-purplish edge, sometimes fringed by a zone of yellowish leaf tissue. On the spots there appear minute brown pycnidia and in artificial culture a perithecial ascus stage has been obtained. This belongs to the genus *Gnomonia.*

LEAF SCORCH

This disease starts as small dark purplish spots like those of leaf spot. As they increase in size, however, they do not develop the white centre. Individual spots may unite to cover a large part of the leaflet and turn dark brown, hence the name " scorch." Infected leaves often turn yellow and the outer ones may wither altogether while the central leaves have few or no spots. Spots may also occur on the fruit stalks and check development of the berries. Tiny black conidial fructifications of the fungus develop on the upper surface of the spots.

Somewhat similar leaf spots, brown with a reddish-purple border and often a yellowish zone outside that, may be caused by another pycnidial fungus, *Septogloeum fragariae* (Lib.) v. Hohnel, which also causes hard rot of the fruit.

GREY MOULD

Many berries can be spoilt and covered by this hairy grey mould. Excessive foliage favours the attack, for which reason it is advisable to avoid heavy manuring v/ith nitrogen. The second and third years' crops are most likely to suffer. On low wet ground the fungus may also attack and kill the root stock and destroy the whole plant. On plants under glass all mouldy berries and any diseased or dying leaves should be picked off. Supporting the fruit on short forked twigs helps it to escape infection. Preventive spraying with Pomarsol has been tried. Out of doors one should avoid too close planting and should spread straw or chaff under the trusses after flowering. The variety Deutsch Evern is said to be particularly susceptible to grey mould.

VERTICILLIUM WILT

Sometimes the outer leaves wither first, turning brown and dry, or develop a brown margin. In other cases the whole plant withers at once and dies quickly. In milder attacks it may linger for a long time and develop a few narrow leaves and weak runners that soon die. Usually there is no discolouration in the root stock until the disease is fairly well advanced, when small dark spots can be found in the flesh. Later on the roots turn dark brown.

One can isolate the fungus from the roots much less readily than from the root stock. Usually one can obtain it from the upper part of the root stock and even in the petiole. *Verticillium* wilt is not the usual cause of black roots. Runners should be taken only from healthy plants and not from those adjacent to wilted ones. It is, indeed, best not to take any runners from beds in which disease is present. Strawberries should not be planted on land known or suspected to contain *Verticillium* from another susceptible crop such as potato, tomato, raspberry or stone fruit.

LEATHERY ROT

In this disease the developing fruit turns brown and, instead of becoming soft and rotten as with grey mould, it becomes tough, leathery and eventually shrivels. As the rot proceeds there is a sharp dividing line between healthy and diseased parts of the berry. The fungus may attack a great many host plants and there is no reason to suppose the late date of its discovery indicates it to have been a recent introduction. Heavy rain favours the disease. Infection comes from the soil and is no doubt spread by splashing during rain. It is therefore apt to be more severe when the beds have not been properly strawed.

A copper lime dust applied 5-6 weeks prior to ripening, followed by a second and third, the last very light to avoid poisoning the fruit, gave best control, but a sulphur dust was also beneficial and less objectionable for use on a soft-fruit crop.

"RUNNING OUT" OR "DECLINE" OF STRAWBERRY CROPS

It is a common experience that after a time strawberries cease to " do" at a particular place and the soil is then often said to have become " tired " of the crop. As in the case of other " tired soils " the phenomenon is not a simple one but may be due to several causes, operating singly or in combination. In the case of strawberries these include cultural factors, virus diseases, root rots, red core disease, and insect pests. Strawberry plants suffering from several diseases and pests tend to develop much the same type of "degeneration," small, weak plants with short-stalked, red-edged leaves, few runners and low cropping capacity.

The roots may be damaged by beetle larvae, especially the June. Where these insects are plentiful the plants wilt and die, especially in June and July.

Nematode infection produces the disease known as red plant. The mites infest the leaves which remain small, brittle and yellowish and in extreme cases turn red. Not only do they weaken the plant but the damage they cause masks the symptoms of virus diseases.

Soil conditions in many places are unsuitable for strawberry cultivation. Very heavy soils should be avoided, so should very light sands, except in areas of high rainfall. Strawberries require humus, and a direct relationship may be traced between the organic-matter content of the soil and the vigour of the plants. Mammal experiments have often shown good results from the use of stable manure. At Long Ashton manuring with waste products of the woollen industry has led to production of high-quality fruit.

If a heavy dressing of farmyard manure has been applied before planting, application of artificial manures seems superfluous, though a little nitrate may be given at the time of blossoming and a little more plus superphosphate after the crop has been picked. When dung is not available strawberries respond to phosphatic fertilizers, especially bone meal or superphosphate, plus a small quantity of nitrate of soda or sulphate of ammonia. Phosphorus-deficient strawberries develop purple and bronze tints in their leaves. Heavy potash manuring is not required except on definitely potassium-deficient soils. If there is chlorine in the fertilizer it may be definitely harmful, so too is sodium. Acid soils of pH 6-7 are said to be best but experience at East Mailing shows that strawberries may do well on soil rich in lime. Presence of lime may, however, lead to minor element deficiencies and excellent results have been obtained by micro-nutrient manuring on limed soils. It may be necessary to supply manganese, iron and copper; Boron deficiency has been demonstrated in sand-culture experiments but not yet in the field.

Methods of cultivation too are not always suited to the requirements of the strawberry plant. Its new roots are formed mainly in late summer and autumn and arise higher up on the root stock than the old ones. If the natural growing conditions of the plant are to be reproduced it should be surrounded with a layer of dead leaves or similar material every year. At Long Ashton good results have followed surface hoeing in August, after which the soil was pressed firmly against the plants. The practice of allowing weeds to grow in the beds after the fruit has been picked and digging them in in autumn is undesirable as the digging destroys the new roots of the strawberries as well as the weeds. Runners take much nourishment from the parent plant and should be removed from a fruiting bed. Runner production should be carried on in separate beds and preferably as a separate industry. It is usually best to pick the blossoms from plants in their first year after planting and to allow them to fruit for the first time in the second year.

A plentiful supply of moisture is necessary for a good crop of fruit and in dry summers irrigation may be a profitable practice. On the other hand ill-drained soils are unsuitable and encourage root rots and red-core disease. Frost may damage many plants, and low-lying areas should not be chosen for strawberry cultivation. The practice of covering the beds with straw helps to protect against frost and drought. To protect the blossoms from late frosts straw should be placed above the plant at night and raked away during the day. It should not be strewn on the ground below the blossoms until frost is no longer to be feared or frost damage may be intensified.

Brown roots. At the lower end of the root stock on any strawberry plant there are always some of last year's roots that have turned blackish-brown. The new roots that arise mainly in autumn higher up on the root stock and from the runners should be white or yellowish throughout, not dark brown or black. A host of fungi is to be found on and in the brown and black roots, many of which are common inhabitants of diseased roots in many plants. Some of them have been shown to be capable of attacking healthy strawberry roots under certain soil conditions. The presence and extent of brown root rot in strawberries depends on the inter-relationship of fungi and soil conditions. In this connection we may mention Hildebrand and Koch's experiments in Canada with the " Wisconsin Tank," an apparatus in which the plants can be grown at a number of controlled temperatures and amounts of soil moisture.

In this way it was demonstrated that there appeared to be no simple relationship between fungal activity and major nutrients, minor nutrients or soil reaction, but that soil moisture and temperature exercised an important effect. At a soil temperature of 45-50° F. and at 60 per cent soil water-holding capacity the strawberry roots remained quite white, but as the temperature rose to 60° F. and moisture content to 80 per cent browning of the roots steadily increased.Brown root rot of strawberries is to be controlled mainly by maintaining good growing conditions with a well-drained soil. Only runners with clean healthy roots should be selected for propagation. In laying out new strawberry beds it is desirable to avoid land that has carried either strawberries or raspberries within the last 4-5 years. Planting in early autumn encourages strong growth.

RED CORE

This is a serious root disease caused by *Phytophthora fragariae* Hickman. The finer rootlets are first destroyed so that the main roots alone remain, looking like "rats' tails" when the root system is washed free from soil. These main roots are then attacked at the tip and the fungus spreads upwards through the vascular cylinder, which it stains reddish-brown, the "red core." The root cortex remains white and healthy for a time but eventually turns black and decays. The fungus has been isolated in pure culture who was able to reproduce the disease by inoculating *P.fragariae* to healthy plants. Red core is worst on soils with high water content but cannot be controlled by draining infected land.

It is probably spread largely by the sale of infected runners and perhaps locally by infected drainage water as the fungus belongs to the group commonly called water moulds.

Research sponsored by the Department of Agriculture for Scotland has led to the breeding and selection of a number of highly resistant strawberry varieties of good commercial quality. These are known as the Auchin-cruive seedlings, from the place where they were bred, the different varieties being distinguished by numbers. No method of soil treatment has proved effective against red-core disease, and the growth of resistant varieties offers the only hope of successful strawberry cultivation on infected land.

CAULIFLOWER 3

CLUB ROOT, OR FINGER AND TOE

In cabbage and its allies the spindle-shaped irregular swellings on the roots are the most obvious symptoms, and are often accompanied by poor development of the head and, in dry weather, by flagging and a bluish tint to the foliage. In swedes, turnips and radishes, where the stem swells to form storage of reserve foodstuff, this is also attacked, and becomes covered with tumours and cracks both above and below soil level. In the swelling the cells of the flesh are enormously enlarged and developed at the expense of the woody tissue so that the tumours easily rot, the decay being completed by bacterial activity. The galls caused on roots by the cabbage gall weevil can easily be distinguished by their more spherical shape and by always having a cavity at the centre, containing the larva. The so-called hybridization nodules on swede roots are also more round, have a more wrinkled surface and never become very large. True club root can always be recognized under the microscope by the gigantic malformed cells of the host. These cells contain innumerable minute spores of the fungus which are liberated into the soil when the swellings decay. They may remain alive in soil for at least 5-6 years.

Living roots of cruciferous plants stimulate the spores to germinate and from each emerges a swarm spore, a naked.speck of protoplasm able to creep and swim through the soil moisture to the root. It may penetrate the latter through wounds or cracks, as where side roots have emerged, but often swarm spores perforate the wall of root hairs and grow and divide within the hair cell.

From the hair the fungus protoplasm penetrates deeper into the root, following especially the cells of the medullary rays between the vascular strands. These thin-walled cells are stimulated to enormous development and cause distortion and disruption of the whole tissue. Within the giant cells many swarm spores unite to form a large mass of fungus protoplasm (plasmodium) which ultimately breaks up into hundreds of tiny resting spores, ready to be set free in the soil by decay of the root.

CLUB ROOT

Infection can take place at temperatures between 43° and 81° F., but from 64-77° F. is the most favourable temperature. This is why late sowings generally suffer much more than early sowings made while the soil is cold. Soil reaction plays a very important part; pH 5.8 is most favourable for spore germination. Swarm spores can emerge from the resting spores at pH 7.5, but are then sluggish. When the soil is packed with spores, however, infection may occur even at pH 8.3. Similarly, experiments have been found that club root infection is distinctly diminished by increasing dressings of lime or marl, but that severe infestations cannot be stopped by liming alone. Sometimes even slight differences in soil reaction have an effect and nitro-chalk gives a healthy crop where sulphate of ammonia gives a diseased one. Sodium, magnesium and barium salts that raise the pH value in the same way as lime have a corresponding effect on the fungus.

Soil moisture is essential for club root infection; if the soil contains only 50 per cent of the water it is able to hold, cabbage can still grow in it, but the swarm spores are unable to move; as the water content rises above this level so does the percentage infection rise in proportion. High soil moisture during winter and heavy spring rains lead to virulent outbreaks. Summer droughts make the plants flag conspicuously so that they are weakened, but infection by the fungus does not increase. Plants from seed sown in August may escape infection in the autumn, but in the following spring, when new roots are produced, these are liable to be attacked. Heavy summer rains may speed up rotting of roots that are already attacked. Club root does not spread in stored produce but if conditions are moist the diseased tissues quickly decay. The sites of turnip or swede pits should never be used for cultivation of a susceptible crop.The type of soil is not very important, except that well-drained soils rich in lime are least prone to club root.

Spread of infection. Soil on which a crop infected with club root has been grown may be regarded as infected for at least 5-6 years. If a cruciferous crop is taken before this time has elapsed, or if many cruciferous ground-keepers and weeds occur, the period of soil infection is correspondingly extended. The sale of soil to gardens, where the purchaser did not know it to be contaminated with club root, judgment being given for the purchaser, so that the seller was obliged to make good the damaged crop. Soil is easily spilt in transport and infection may thus be spread to other areas. Infected material may also be carried by tools, on horses' hoofs, tractor treads and the boots of workers. It seems that it may be also spread to hitherto uninfected areas by particles of soil blown by the wind. Infection may also be water-borne, as when diseased turnips have been thrown into a pit which overflowed as a result of heavy rains.

Infection in farmyard manure is a very serious problem. Observations indicate that the spores may pass through the intestines of ruminants in a living condition, but this is probably not of practical importance. The damage is done by contamination of stable manure with waste infected fodder such as turnips. The sediment of liquid manure may also be infectious. Hence bought stable manure and kitchen refuse are to be mistrusted, but on the whole consistent use of heavy dressings of non-infected stable manure seems to lead to less club root. Cabbage plants frequently become infected in the seed bed, and are sold when bearing only a slight swelling easily overlooked. These are a frequent source of infection in private gardens. The seed itself cannot be infected, but infection may be present in soil particles in a seed sample.

CANKER

Phoma lingam may attack seedlings, half-grown plants, cabbage heads in storage, seed-bearing plants and the seed itself. It is one of the causes of damping off in cabbage, broccoli, and turnip seedlings. The plants rot at soil level, and when the disease was caused by *P. lingam* small dark pycnidia appear on them, out of which, under moist conditions, ooze amethyst-coloured drops containing the spores *P. lingam* infection at this stage comes from infected seed.These seedlings attacked by the primary seed-borne phase of the disease almost always die, but the disease spreads from them to other plants in the seed-bed. These when planted out develop the canker phase. In this the stem is usually invaded just below soil level. It turns black and decays so that the plant becomes loose in the soil and may eventually fall over.

Secondary roots may be produced from the stem above the canker and keep the plant alive for some time. The leaves often take on a blue or purplish hue, but similar colouration may result from other injuries to the root system as by larvae of the cabbage fly. Pycnidia form on the rotting tissue or at the edge of the canker next to the healthy stem.The disease may progress further in cabbage heads stored for the winter. The fungus grows up through the stump into the leaves, which turn black from the base upwards and outwards. They become loosened from the stem so as to produce a curious crater-like hollow round it. Ultimately pycnidia appear in the diseased tissue.

If there are a few infected plants in a seed crop the disease may easily spread considerably on the ripening pods. *Phoma lingam* infection is often confused with " dark spot" caused by *Alternaria brassicae* and *A. oleracea,* but it is of great importance because it leads to production of infected seed. In a seeding crop numerous dark spots appear on leaves, stems and pods and on these the pycnidia are gradually produced. From the spots on the pods the fungus penetrates the seeds where it occurs as resting mycelium between the embryo and the seed coat. Thus infection of the seed takes place between flowering and seed ripening, but the seedling becomes infected during germination. It is not possible to tell whether a seed sample is infected or not merely by inspecting it, though some infected seeds may be lighter, greyish and somewhat shrivelled. These would float if the sample were shaken up in water. If a germination test is made, however, the diseased seeds can be detected by the pycnidia which appear on them and on the seedlings grown from them. Such a test takes from 3 to 4 weeks.

The same fungus causes a disease of swedes called dry rot. On that host it has been named *Phoma napobrassicae* Rostr. Numerous other names have been given to it, such as *P. oleracea* Sacc., P. *brassicae* Thiim, P. *siliquastrum* Desm. and *Aposphaeria brassicae* Thiim., but they are all superseded by the oldest, viz. *Phoma lingam* (Tode) Desm.

RING SPOT

Almost all above-ground parts of the plant may be affected. On the leaves the spots are usually numerous, round, and light brown to grey-brown with a narrow olive-green border. They bear innumerable small pycnidia or perithecia, usually arranged in concentric rings, whence the name ring spot. The fructifications may be so numerous that the whole spot looks black. In severe attacks the outer leaves of cabbage turn yellow and fall. Spots on the outer leaves of a cabbage head may be so numerous that from a short distance it resembles a red cabbage. Similar spots occur on stems and pods of plants grown for seed. Whereas pods spotted by *Alternaria* tend to open prematurely, those attacked by ring spot are apt to remain closed even when the seed is ripe.

Phoma lingam, the cause of cabbage canker, may be responsible for somewhat similar spots, but there is no connection between it and the true ring spot fungus. *P. lingam* spots bear far fewer pycnidia. The ring spot perithecia form during autumn and ripen about December. Recent investigations have shown that the spores produced by the pycnidia are not conidia and are unable to infect leaves or to grow in artificial culture. Presumably they are male cells or spermatia, analogous to those produced by rust fungi, but their precise function remains unknown. Ascospore germination is slow at 36° F., but rapid at temperatures between 60° and 72° F. At the latter temperature infection may take place in about a day, but there is an incubation period of 2 to 3 weeks before the spots are clearly developed. The fungus is believed to survive from one crop to the next on old leaves, stems and seed pods. Infection is mainly by ascospores blown by the wind.

DOWNY MILDEW

The upper surface of the leaf bears small, pale-yellow, angular spots, often speckled with brown, which may grow together to form irregular patches. On the under-surface the corresponding areas are covered with a light-grey mould formed by the much-branched conidiophores and conidia of the fungus. The disease is commonest on seedlings in the seed bed and on the lower leaves of young plants, but it may occur on older plants and even on stored cabbage heads.

In the latter circumstances the infection came through the stalk from infected lower leaves of the mature cabbage. The pith turns dark grey, and grey lesions with a yellow-brown margin occur on the leaves throughout the head. Downy mildew may also spoil the sprouts in brussels sprouts.

The downy mildew affecting most cruciferous plants, cabbage and its allies, turnip, swede, seakale, mustard, radish, stock, wallflower and weeds like shepherd's purse and bitter cress, has traditionally been regarded as a single species, *Peronospora parasitica* Tul. This name is still used for it in British literature. In Switzerland, however, Gaumann has shown that *P. parasitica* in this wide sense includes a considerable number of races specialized on particular host plants and these he prefers to call different species. In his nomenclature the downy mildew of cabbage and its allies becomes *P. brassicae* Gaumann. Because of this degree of specialization there is no risk of cabbage being infected from, the downy mildew so common on shepherd's purse.

Within the tissue of the diseased leaves the fungus forms its sexual reproductive bodies, oospores, which can remain alive in the decaying leaves in soil for several years. They are, however, not very abundantly produced and the fungus probably overwinters mostly in the form of mycelium in living plants. Laubert mentions the possibility of the mycelium growing into the seed pods and seeds, but seed-borne infection is not usually regarded as important in this disease. The fungus can grow and form conidia at a temperature as low as 40° F. and the spores germinate readily in water between 46° and 54° F., less freely at 66° F. and scarcely at all at 73-77° F.

Importance. Downy mildew does most damage to plants indoors and in frames. Only occasionally under moist conditions does it become important out of doors. Plants infected in the seedling stage usually grow away from the disease. It has damaged heads of savoys particularly by persisting through the summer on the lower leaves, attacking the outer leaves of the head and growing in from there to cause brown spotting of the inner leaves in stored heads. Kale seems less susceptible than cabbage.

Soil in which cabbages or their allies have been grown during the last few years should not be used for another cabbage crop. Development of the fungus is checked by dry air, hence the plants should be kept dry and the atmospheric humidity low by giving plenty of ventilation to seed beds in frames and glasshouses. Spraying with Bordeaux mixture and other copper sprays also definitely checks downy mildew, especially when the under-side of the leaves is wetted, but spraying should be done early in the day and in weather when the moisture will dry quickly. Copper lime dusts may also be helpful, sulphur dusts are less effective.

WHITE BLISTER

This disease is also sometimes called white rust, but that is a misleading name as the fungus which causes it is allied to the downy mildews and not to the true rust fungi. It is found on almost all members of the family *Cructferae:* turnip, swede, cabbage, and its allies, mustard, cress, horseradish, arabis, honesty, wallflower, and numerous weeds, especially shepherd's purse. Any above-ground part of the plant may be attacked and in flowering plants the stems are often twisted, contorted and much swollen. The spores are formed in large white blisters under the epidermis, which burst open to expose powdery white pustules.

DARK LEAF SPOT

The dark leaf spot fungi are commonly seed-borne and may attack seedlings, causing brown spots and streaks on stems and leaves. This phase of the attack is of minor importance in western Europe, but spotting of mature leaves and especially of the flower stalks, flowers and pods of seed crops may be of some significance. Two related fungi are concerned in the disease, which differ mainly by the size of their spores and other microscopic characters. Both cause rounded spots up to an inch across on leaves.

The spots are pale yellow with a yellow margin on green leaves and green on red cabbage leaves, best seen on the upper side of the leaf. In damp weather the upper surface, and to some extent the lower also, becomes covered by conidiophores and conidia.

In *A. brassicae* they are pear-shaped with long beaks and form a pale-greyish felt-like coating; in *A. oleracea* there is a dark olive-green, closely velvety growth, bearing the conidia in concentric rings. The spots can be found on leaves all the year round. In seed crops, when the weather is damp and warm, these fungi spread rapidly over the stems after flowering and form pale oblong spots, soon covered with a dark layer of conidia, on the ripening pods. In a few days of favourable weather the appearance of a whole seed bed may be changed from green to brown or almost black.

On the pods the mycelium grows through the wall into the seeds. Severely attacked pods shrivel and die. In others the seeds ripen but are weak and germinate with only a weak shoot soon killed by the fungus. Slightly infected seeds can be protected by seed disinfection and will in any case give a vigourous seedling if conditions favour rapid germination.

Presence of these moulds causes the pods to split open prematurely so that much seed is lost by shedding and the yield obtained from the crop is reduced. The quality of the seed is also adversely affected. Spots on the leaves are seldom of much importance in themselves, but help the fungi to persist in the crop all the year round. The lowest leaves of a young seed crop are usually infected. They wither in the spring and on them are produced enormous numbers of conidia by which the flowering and fruiting stems become infected later on.

GREY MOULD

Cabbage heads, especially when stored, may be attacked by the common grey mould, which causes watersoaked grey-green to brownish patches on the leaves. These become covered with the familiar grey coating of conidiophores and conidia and eventually irregular black sclerotia form in the rotting tissue. Grey mould thrives under damp conditions, especially at a moderate temperature, and may cause great loss to cabbages in storage.

It is important that the cabbages be dry and well aired for some days before being stored, and that care be taken to avoid bruising them, as the fungus easily gains entry to the tissues through wounds. Hearts of spring cabbage may also be attacked in the field, especially after damage by frost. *Botrytis cinerea* may also damage brussels sprouts even out of doors during the winter months. Early Dwarf and Long Island are particularly susceptible to it. Early varieties generally suffer most.

SCLEROTINIA ROT

In cabbage this is principally a disease of the stored heads. Larger or smaller parts of them turn black and slimy and on these the white, cottony mycelium characteristic of the fungus develops. Associated with it are large sclerotia, at first white, then with a black rind. Cabbages should be dry when stored and should be handled carefully so as not to be bruised, for damaged tissue forms an easy means of access for this fungus and for *Botrytis cinerea,* etc.

The disease is worst when the straw in the pit becomes damp. Cabbage varieties differ in susceptibility to *Sclerotinia* rot; of red cabbages Amager and spring red cabbage are resistant, and of ordinary cabbages high and half-high Amager.

BACTERIAL LEAF SPOT

This disease, has been found on cabbage, cauliflower, broccoli, savoy and brussels sprouts. Numerous small watersoaked spots appear on the leaves and soon turn brownish or grey with translucent edges. At first they are round but become angular or so by the fine veins of the leaf. If they are very numerous several may coalesce and the leaves may become ragged and torn. In severe infections the leaves may turn yellow and fall. The causal bacterium enters leaves through the stomata and can also attack flowers and the curd of cauliflower heads. It grows best at 77° F., not at all at temperatures over 86° F. and is favoured by high humidity. Seed disinfection with corrosive sublimate may help to prevent infection, and spraying with Bordeaux mixture plus a spreading agent may check its spread.

BLACK ROT

Black rot may affect almost any cultivated cruciferous crop and some wild species as well, especially cabbage, red cabbage, savoy, brussels sprouts, broccoli, cauliflower, kohlrabi, swede, turnip, mustard, charlock, radish, wallflower and stock.

Plants may be attacked at any age but especially when nearing maturity. In cabbage a larger or smaller section of a leaf typically turns yellow and bears a network of blackened veins. This blackening spreads down the vascular bundles, to the upper part of the stem. Especially in turnip, infection may also start in the roots, which have been damaged by larvae of the cabbage fly and other insects. In severe attacks so many of the outer leaves are destroyed that the cabbage plant appears " leggy " and its head becomes lopsided and loose-leaved. The blackened vascular bundles may appear in the head leaves too.

When infection comes from the seed the cotyledons become infected 12-20 days after sowing. Many plants may be killed in the seed-bed or, if they survive, spread the disease when they are planted out. The rot caused by *X. campestris* alone is rather dry and limited in extent but it may open the way to secondary organisms which set up a rapid evil-smelling soft rot.

Infection may come from the seed or the soil, especially if diseased cabbage or turnip heads, roots or leaves have come in with manure. From diseased plants and from infected soil the bacteria are spread by rain splashes, also by snails, by the larva; of insects like the cabbage fly and by workers among the plants. Most infections occur through the water pores on the edge of the leaves. These exude drops of water during the night and if bacteria get into these drops they multiply and enter the terminal strand of the vascular system below the water pore. The bacteria can grow at temperatures between 40 and 100° F. They thrive on cabbage which has heated in a silo. Freezing and thawing does not kill the bacteria, which are able to remain alive in the soil from one year to the next. Nor are they killed by drying and survive for one year, possibly for three years, in stored seed. Black rot seems to occur throughout the world wherever members of the cabbage family are grown.

SOFT ROT

Plant organs that are used for the storage of foodstuffs are often affected by a soft stinking rot caused by bacteria. Probably the commonest cause of the trouble is *Bacterium carotovorum* which has received several other names in the past according to the host plant in which it was found, e.g. *Bacillus oleraceae* Harr., *B. omnivorus* Van Hall, *B. apiovorus* Worm, and *B. hyacinthi septicus* Heinz. Closely allied species of soft-rotting bacteria are *Bacterium aroideae* (Towns.) Stapp, which can attack arum corms, iris leaves, potato tubers, etc., and *Bacterium phytophthorum* (Appel) Burgwitz, the cause of black-leg of potato.

Host plants. Soft rot is most frequently found in mature turnip roots, next in carrots and quite often in radish, swede, cabbage and its allies, potato tubers, especially in wet, warm soil, celery, seakale, parsnip, onion, leek, cucumber, melon, horseradish, tomato, capsicum, egg plant, arum, iris, hyacinth, cyclamen, and probably parsley, wallflower and violet.It is always soft and fleshy organs that are attacked, usually storage organs, like carrot and turnip roots, melon fruits, iris rhizomes, but sometimes other organs like the soft stems of cucumber and melon and the fleshy leaf bases of iris. It is characteristic of soft rot that the tissues dissolve into a sloshy mess unable to support their own weight. The rotting organ therefore collapses while watery drops and slime containing swarms of bacteria ooze out. There is often a peculiar smell such as one associate with rotten cabbage.

Bacterium carotovorum is able to move freely in water by means of its flagellae. It is emphatically a wound parasite but once it has obtained entrance to plant tissues it secretes a poison which diffuses into neighbouring healthy cells and kills them. The middle lamella, the substance between plant cells which binds them together, becomes dissolved, so that the tissue becomes a structureless mass. Usually there is no change in colour but sometimes there is a slight brown discolouration. The bacteria multiply most rapidly at 80-86° F., hence their activity is restricted by cold storage.Rotting at the neck in swedes and turnips similarly often follows injury by the swede midge. The heart leaves of celery are often destroyed by this organism.

BROCCOLI MOSAIC

The first symptom of infection in broccoli is a clearing of the finer leaf veins which appears first towards the base of young leaves. It may become more obvious as the leaf develops so that the veins become quite yellow. This condition is slowly replaced by vein banding, in which a dark-green band of tissue follows the larger veins and the intervening portions of the leaf blade become pale or yellowish. Small dead spots, at first white, then pale brown, may appear in the yellow areas. The leaves may be twisted and deformed and there is a tendency for the outer leaves to bend backwards and downwards instead of arching over the curd. In severe infections the whole plant is stunted and no marketable curd is produced.

CABBAGE BLACK RING SPOT

This disease is common on cabbage and is easily recognized by the small black ring-like markings which occur scattered over the leaves of infected plants. On cauliflower the virus causes a fine light-green mottle. The virus also attacks a number of ornamental plants such as arabis, wallflower, hesperis, and stock, and causes "breaking" in the colour of self-coloured flowers. It is transmitted by aphides.

BROWNING OF CAULIFLOWER

Boron deficiency on cauliflower is widespread in some districts. Small transparent areas appear in the pith of the stem and on the surface of the maturing head. These turn brown or even almost black. Aifected heads do not decay but are very bitter and unpalatable. In severe boron deficiency the leaves may be narrow and the formation of a head may be delayed or even prevented. The disease can be controlled by manuring with borax at the rate of 5 to 25 Ib. per acre according to the severity of the trouble.

The usual symptoms are watersoaked spots in the pith of the main stem. These turn brown and may break down into cavities. In extreme cases the oldest leaves may be mottled at the edges, be curled and irregular in shape and develop cross-cracking on the upper surface of the leaf stalk. Early maturing varieties are most prone to the trouble.

It seems that when potassium is deficient cabbage and all its allies develop pale leaf edges which later shrivel, also, sometimes dead spots on the leaf blade and brown streaks on the leaf stalks. The heads may be loose and defective, brussels sprouts develop few sprouts, there are green leaves in cauliflower heads and red cabbage leaves turn greenish. Potassium-deficient plants seem more susceptible to aphis and downy mildew than normal plants.

WHIPTAIL

Whiptail is a condition in which the leaves, usually of cauliflowers, are narrow, irregular and wavy and in extreme cases consist merely of a bare midrib or of a midrib with irregular lodes of leaf blade on each side of it. The condition has been reproduced artificially by growing cauliflower and broccoli plants in sand cultures lacking molybdenum. Whiptail usually occurs on acid soils very deficient in lime, but liming alone does not always cure it.

Molybdenum in the soil is believed to be more available to plants in alkaline or limed soils than in acid ones. The disease has been cured by manuring with very small quantities of sodium or ammonium molybdate.

BLACK SPOT OF CABBAGES

There were complaints from several places that some of the leaves in mature cabbage heads were covered with small black spots. The appearance of the diseased cells resembled that of cells in turnips affected with brown heart, and boron deficiency may have been involved in the cabbage trouble. Similar black specks in cabbage and cauliflower leaves may arise in transit and during storage. They have been produced experimentally by keeping cabbages at room temperature in an atmosphere with little or no oxygen and by keeping them in a well-ventilated room at temperatures near 32° F. for 2-3 weeks. Black spot has also been found in the field after a sharp fall in temperature

SPLITTING OF THE HEAD IN CABBAGE

In very wet weather the roots absorb so much water that growth of the inner juicy tissues of the cabbage head becomes too rapid and the outer tissues are burst open. This is especially likely to happen when heavy rain follows a drought. By driving a spade into the soil on both sides of the plant or by twisting it half round in the soil a considerable number of roots will be severed and excess absorption of water will be prevented.

MAIZE

SMUT

Maize smut may attack all the young parts of the plant above ground, even the adventitious roots. It is least prevalent in the leaves. Large tumours are formed, sometimes as much as 7 inches long by 4 inches broad, usually on the lower parts of the stem. To a lesser extent both male and female inflorescences may be partially replaced by irregular outgrowths full of black smut spores. The shape of the plant may indicate infection with smut, for infected flower heads become so heavy that they bend down in some unnatural position. When the stem is attacked deformations may arise both above and below the smutted area. The smut tumours may assume various peculiar shapes. At first they are covered with a grey, or whitish, shining skin made up partly of plant tissue, partly of a membrane formed by the fungus. Until it ripens the blackish-brown spore mass inside is moist and greasy but it rapidly changes to a dry powder which spreads the disease far and wide.

The infected plants and parts of plants should be removed and burnt as soon as they are seen and all debris of the crop should be destroyed in the autumn. Infection is spread by maize debris in manure. It is stated abroad that the fungus is not carried by the seed, but its appearance in localities where maize has never been grown before indicates that this does occur to some extent. Possibly it is carried only on the surface of the seed. Attempts to breed resistant varieties are in progress in countries where maize is an important crop.

4

Vegetable Precise Processing Skills

VEGETABLES VARIETIES

Vegetable processors must appreciate the substantial differences that varieties of a given vegetable will possess. In addition to variety and genetic strain differences with respect to weather, insect and disease resistance, varieties of a given vegetable will differ in size, shape, time of maturity, and resistance to physical damage.

Varietal differences then further extend into warehouse storage stability, and suitability for such processing methods as canning, freezing, pickling or drying. A variety of peas that is suitable for canning may be quite unsatisfactory for freezing and varieties of potatoes that are preferred for freezing may be less satisfactory for drying or potato chip manufacture. This should be expected since different varieties of a given vegetable will vary somewhat in chemical composition, cellular structure and biological activity of their enzyme system.

HARVESTING AND PRE-PROCESSING

When vegetables are maturing in the field they are changing from day to day. There is a time when the vegetable will be at peak quality from the standpoint of colour, texture and flavour.This peak quality is quick in passing and may last only a day. Harvesting and processing of several vegetables, including tomatoes, corn and peas are rigidly scheduled to capture this peak quality. After the vegetable is harvested it may quickly pass beyond the peak quality condition.

This is independent of microbiological spoilage; these main deteriorations are related to:

a. Loss of sugars due to their consumption during respiration or their conversion to starch; losses are slower under refrigeration but there is still a great change in vegetable sweetness and freshness of flavour within 2 or 3 days;

b. Production of heat when large stockpiles of vegetables are transported or held prior to processing.

At room temperature some vegetables will liberate heat at a rate of 127,000 kJ/ton/day; this is enough for each ton of vegetables to melt 363 kg of ice per day. Since the heat further deteriorates the vegetables and speeds micro-organisms growth, the harvested vegetables must be cooled if not processed immediately.

But cooling only slows down the rate of deterioration, it does not prevent it, and vegetables differ in their resistance to cold storage. Each vegetable has its optimum cold storage temperature which may be between about 0-100 C (32-500 F).

c. The continual loss of water by harvested vegetables due to transpiration, respiration and physical drying of cut surfaces results in wilting of leafy vegetables, loss of plumpness of fleshy vegetables and loss of weight of both.

Moisture loss cannot be completely and effectively prevented by hermetic packaging. This was tried with plastic bags for fresh vegetables in supermarkets but the bags became moisture fogged, and deterioration of certain vegetables was accelerated because of buildup of CO2 and decrease of oxygen in the package. It therefore is common to perforate such bags to prevent these defects as well as to minimise high humidity in the package which would encourage microbial growth.

Shippers of fresh vegetables and vegetable processors, whether they can, freeze, dehydrate, or manufacture soups or ketchup, appreciate the instability and perishability of vegetables and so do everything they can to minimise delays in processing of the fresh product. In many processing plants it is common practice to process vegetables immediately as they are received from the field.

To ensure a steady supply of top quality produce during the harvesting period the large food processors will employ trained field men; they will advise on growing practices and on spacing of plantings so that vegetables will mature and can be harvested in rhythm with the processing plant capabilities. This minimises stockpiling and need for storage.

Cooling of vegetables in the field is common practice in some areas. Liquid nitrogen-cooled trucks may next provide transportation of fresh produce to the processing plant or directly to market. Upon arrival of vegetables at the processing centre the usual operations of cleaning, grading, peeling, cutting and the like are performed using a moderate amount of equipment but a good deal of hand labour also still remains.

RECEPTION

This covers qualitative and quantitative control of delivered vegetables. The organoleptic control and the evaluation of the sanitary state, even if they are very important steps in vegetables' characteristics assessment, cannot establish their technological value.

On the other hand, laboratory controls do not precisely establish their technological properties because of the difficulty in putting into showing some deterioration when using rapid control methods.

One correct method of vegetable quality appraisal is their overall evaluation based on the whole complex of data that can be obtained by combining an extensive organoleptic evaluation with simple analysis that can be performed rapidly in plant laboratory. These analysis can be:

a. Refractometric extract (tomatoes, fruit, etc.);
b. Specific weight (potatoes, peas, etc.);
c. Consistency (measured with tenderometres, penetrometres, etc.);
d. boiling tests, etc.

TEMPORARY STORAGE

This step should be as short as possible and better completely eliminated. Vegetables can be stored in:

a. Simple stores, without artificial cooling;
b. In refrigerated stores; or, in some cases,
c. In silos (potatoes, etc.).

Simple stores should be covered, fairly cool, dry and well ventilated but without forced air circulation which can induce significant losses in weight through intensive water evaporation; air relative humidity should be at about 70-80%.

Refrigerated storage is always preferable and in all cases a processing centre needs a cold room for this purpose, adapted in volume I capacity to the types and quantities of vegetables (and fruits) that are further processed.

WASHING

Washing is used not only to remove field soil and surface micro-organisms but also to remove fungicides, insecticides and other pesticides, since there are laws specifying maximum levels of these materials that may be retained on the vegetable; and in most cases the allowable residual level is virtually zero. Washing water contains detergents or other sanitisers that can essentially completely remove these residues.

The washing equipment, like all equipment subsequently used, will depend upon the size, shape and fragility of the particular kind of vegetable:

- Flotation cleaner for peas and other small vegetables;
- Rotary washer in which vegetables are tumbled while they are sprayed with jets of water; this type of washer should not be used to clean fragile vegetables;

SORTING

This step covers two separate operations:

a. Removal of non-standard vegetables (and fruit) and possible foreign bodies remaining after washing;
b. Quality grading based on variety, dimensional, organoleptical and maturity stage criterion.

SKIN REMOVAL/PEELING

Some vegetables require skin removal. This can be done in various ways:

a. *Mechanical*: This type of operation is performed with various types of equipment which depend upon the result expected and the characteristics of the fruit and vegetables, for example:
    i. A machine with abrasion device (potatoes, root vegetables);
    ii. Equipment with knives (apples, pears, potatoes, etc.);
    iii. Equipment with rotating sieve drums (root vegetables). Sometimes this operation is simultaneous with washing (potatoes, carrots, etc.) or preceded by blanching (carrots).
b. *Chemical*: Skins can be softened from the underlying tissues by submerging vegetables in hot alkali solution. Lye may be used at a concentration of about 0.5-3%, at about 93° C (2000 F) for a short time period (0.5-3 min). The vegetables with loosened skins are then conveyed under high velocity jets of water which wash away the skins and residual lye.

    In order to avoid enzymatic browning, this chemical peeling is followed by a short boiling in water or an immersion in diluted citric acid solutions.

    It is more difficult to peel potatoes with this method because it is necessary to dissolve the cutin and this requires more concentrated lye solutions, up to 10%.
c. Thermal

WET HEAT (STEAM)

Other vegetables with thick skins such as beets, potatoes, carrots and sweet potatoes may be peeled with steam under pressure (about 10 at) as they pass through cylindrical vessels. This softens the skin and the underlying tissue. When the pressure is suddenly released, steam under the skin expands and causes the skin to puff and crack. The skins are then washed away with jets of water at high pressure (up to 12 at).

DRY HEAT (FLAME)

Other vegetables such as onions and peppers are best skinned by exposing them to direct flame (about 1 min at 1000° C) or to hot gases in rotary tube flame peelers. Here too, heat causes steam to develop under skins and puff them so that they can be washed away with water.

Manual peeling only use when the other methods are impossible or sometimes as a completion of the other three ways. Average losses at this step are given in Table.

Table: Losses at Vegetable Peeling, in %

	Peeling methods		
Vegetables	Manual	Mechanical	Chemical
Potatoes	15-19	18-28	-
Carrots	13-15	16-18	8-10
Beets	1416	13-15	9-10

SIZE REDUCTION

This step is applied according to specific vegetable and processing technology requirements.

BLANCHING

The special heat treatment to inactivate enzymes is known as blanching. Blanching is not indiscriminate heating. Too little is ineffective, and too much damages the vegetables by excessive cooking, especially where the fresh character of the vegetable is subsequently to be preserved by processing. This heat treatment is applied according to and depends upon the specificity of vegetables, the objectives that are followed and the subsequent processing/ preservation methods.Two of the more heat resistant enzymes important in vegetables are catalase and peroxidase. If these are destroyed then the other significant enzymes in vegetables also will have been inactivated. The heat treatment to destroy catalase and peroxidase in different vegetables are known, and sensitive chemical tests have been developed to detect the amounts of these enzymes that might survive a blanching treatment.

Because various types of vegetables differ in size, shape, heat conductivity, and the natural levels of their enzymes, blanching treatments have to be established on an experimental basis. As with sterilisation of foods in cans, the larger the food item the longer it takes for heat to reach the centre. Small vegetables may be adequately blanched in boiling water in a minute or two, large vegetables may require several minutes. Blanching as a unit operation is a short time heating in water at temperatures of 100° C or below. Water blanching may be performed in double bottom kettles, in special baths with conveyor belts or in modern continuous blanching equipment.

In order to reduce losses of hydrosoluble substances (mineral salts, vitamins, sugars, etc.) occurring during water blanching, several methods have been developed:

- Temperature setting at 85-95° C instead of 100° C;
- Blanching time has to be just sufficient to inactivate enzymes catalase and peroxidase;

- Assure elimination of air from tissues.

An illustration of blanching parametres is seen in Table given below.

Table. Blanching Parametres for Some Vegetables

Vegetables	Temperature, °C	Time, min.
Peas	85-90	2-7
Green beans	90-95	2-5
Cauliflower	Boiling	2
Carrots	90	3-5
Peppers	90	3

Steam heat treatment can also be applied instead of water blanching as a preliminary step before freezing or drying, as long as the preservation method is only used for enzyme inactivation and not to modify consistency.

For drying, the vegetables are conveyed directly from steaming equipment to drying installations without cooling. Vegetable steaming is carried out in continuous installations with conveyer belts made from metallic sieves.

Cooling of vegetables after water blanching or steaming is performed in order to avoid excessive softening of the tissues and has to follow immediately after these operations; one exception is the case of vegetables for drying which can be transferred directly to drying equipment without cooling.

Natural cooling is not recommended because is too long and generates significant losses in vitamin C content. Cooling in pre-cooled air (from special installations) is sometimes used for vegetables that will be frozen

Cooling in water can be achieved by sprays or by immersion; in any case the vegetables have to reach a temperature value under 37° C as soon as possible. Too long a cooling time generates supplementary losses in valuable hydrosoluble substances; in order to avoid this, the temperature of the cooling water has to be as low as possible.

CANNING

Large quantities of vegetable products are canned. A typical flow sheet for a vegetable canning operation (which also applies to fruit for the most part) covers some food process unit operations performed in sequence: harvesting; receiving; washing; grading; heat blanching; peeling and coring; can filling; removal of air under vacuum; sealing/closing, retorting/heat treatment; cooling; labelling and packing. The vegetable may be canned whole, diced, puréed, as juice and so on.

ON-LINE SIMPLIFIED METHODS FOR ENZYME ACTIVITY CHECK

Peroxidase test:

a. Solutions. In order to check the peroxidase activity two solutions have to be prepared:
   - 1% guaiacol in alcohol solution (1 g guaiacol is dissolved in about

50 cm^3 of 96% ethylic alcohol and then this preparation is brought to 100 cm° with the same solvent);
- Peroxide solution 0.3% (1 cm^3 perhydrol is brought to 100 cm^3 with distilled water.

b. Sampling. From various parts of the material samples are taken (about 20-30 pieces, etc.); the material is then crushed in a laboratory bowl in order to obtain an average sample.
c. Check. Prom the average sample, 10-20 g of material is taken in a medium capacity test tube; on this sample are poured: 20 cm^3 distilled water; 1 cm^3 of 1% guaiacol solution; 1.6 cm^3 of peroxide solution.

The contents of the test tube is shaken well. The gradual appearance of a weak pink colour indicates an incomplete peroxidase inactivation - reaction slightly positive. If there are no tissue colour modifications after 5 minutes, the reaction is negative and the enzymes have been inactivated. As an orientative check it is also possible to simply pour a few drops of 1% guaiacol solution and 0.3% peroxide solution directly on blanched and crushed vegetables. A rapid and intensive brown-reddish tissue colouring indicates a high peroxidase activity (positive reaction).

CATALASE TEST

In order to identify the catalase enzyme activity, 2 g of dehydrated vegetables are well crushed and mixed with about 20 cm^3 of distilled water. After 15 min softening, 0.5 cm^3 of a 0.5% or 1% peroxide solution is poured on prepared vegetables. In the presence of catalase, a strong oxygen generation is observed for about 2-3 minutes. These tests are of a paramount importance in order to determine the vegetable blanching treatments (temperature and time); incomplete enzyme inactivation has a negative effect on finished product quality.

For cabbage catalase inactivation by blanching is sufficient; blanching further to peroxidase inactivation would have negative effects on product quality and even complete browning. For all other vegetables and for potatoes, both tests MUST be negative, for catalase and for peroxidase.

FRESH VEGETABLE STORAGE

The vegetables can be stored, in some specific natural conditions, in fresh state, that is without significant modifications of their initial organoleptic properties. Fresh vegetable storage can be short term; this was briefly covered under temporary storage before processing. Also fresh vegetable storage can be long term during the cold season in some countries and in this case it is an important method for vegetable preservation in the natural state.In order to assure preservation in long term storage, it is necessary to reduce respiration and transpiration intensity to a minimum possible; this can be achieved by:

a. Maintenance of as low a temperature as possible (down to 0° C),
b. Air relative humidity increased up to 85-95 % and
c. CO_2 percentage in air related to the vegetable species.

Vegetables for storage must conform to following conditions: they must be of one of the autumn or winter type variety; be at edible maturity without going past this stage; be harvested during dry days; be protected from rain, sun heat or wind; be in a sound state and clean from soil; be undamaged.

From the time of harvest and during all the period of their storage vegetables are subject to respiration and transpiration and this is on account of their reserve substances and water content. The more the intensity of these two natural processes are reduced, the longer sound storage time will be and the more losses will be reduced.

For this reason, vegetables have to be handled and transported as soon as possible in the storage conditions (optimal temperature and air relative humidity for the given species). Even in these optimal conditions storage will generate losses in weight which are variable and depend upon the species. Some optimal storage conditions are shown in table given below.

Table. Optimal conditions for fresh vegetable storage

Vegetables	Storage conditions	
	Temperature, °C	**Relative humidity, %**
Potatoes	+1...+3	85-90
Carrots	0 ... +1	90-95
Onions	0 ... +1	75-85
Leeks	0 ... +0.5	85-90
Cabbage	-1 ... 0	90-97
Garlic	0 ... +1	85-90
Beets	0 ... +1	90-95

SELECTION AND STORAGE

It is important to select potatoes of high specific gravity since this characteristic indicates superior yield and lower oil absorption. It is even more important to select potatoes with low reducing sugar contents or to store them at temperatures conducive to the minimising of these substances. Sprouting and fungal damage must also be minimised by the storage conditions.

PEELING

The ideal peeling operation should only remove a very thin outer layer of the potato, leaving no eyes, blemishes, or other material for later removal by hand trimming. It should not significantly change the physical or chemical characteristics of the remaining tissue. Preferably peeling should use small amount of water and result in minimal effluent; compromises will have to be made in all of these aspects of peeling.

First, the potatoes are thoroughly washed, not only for sanitary reasons, but also to prevent dirt of grit from abrading the equipment the tubers will dater contact. Washing may take place in streams, as the potatoes are being conveyed by water streams, or in equipment provided with means for scrubbing the potato with brushes or rubber rolls.

In barrel-type washers, potatoes are cleaned by being tumbled and rubbed against each other and against the sides of the barrel while they are immersed in, or sprayed with, water. After washing, the potatoes are allowed to drain, usually on mesh conveyors, and they travel over an inspection belt where foreign material and defective tubers are removed. The more common peeling methods are abrasion, lye immersion, and steam.

Abrasion peelers which may be either batch or continuous, use disks or rollers coated with grit to grind away the potato surface. An important design feature is to ensure that all surfaces of the tuber are equally exposed to the rasping action. The peel fragments are flushed out of the unit by water sprays.

Such systems work best with uniform, round, undamaged potatoes. Some of the advantages of abrasion peelers are their simplicity, compactness, low cost, and convenience. They are particularly suitable for peeling potatoes intended for chipping, since they do not chemically alter the surface layers. About 10% of the original tuber weight is lost through abrasion peeling prior to chipping.

SLICING

The peeled potatoes are cut into slices from 1/15 to 1/25 in. by rotary slicers. Centrifugal force presses the tuber against stationary gauging shoes and knives. Thickness is varied, not only to meet consumer preferences, but also to fit the condition of the tubers and the frying temperature and time.

Slices produced at any one time must be very uniform in thickness, however, in order to obtain uniformly coloured chips. Slices with rough or torn surfaces lose excess solubles from ruptured cells and absorb larger amounts of fat.

It is necessary to remove the starch and other material released from the cut cells from the surface of slices so that the slices will separate readily and completely during frying. The slices are washed in stainless steel wire-mesh cylinders or drums rotating in a rectangular stainless steel tank. After washing and an additional rinse in similar equipment, the potatoes may or may not be dried.

FRYING

The capacity of the fryer is generally the limiting factor in the process line. Most manufacturers currently use continuous fryers but some batch equipment is still employed.

Modern continuous fryers have the following essential elements:

1. A tank of hot oil in which the chips are cooked;
2. A means for heating and circulating the oil;
3. A filter for removing particles from oil;
4. A conveyor to carry chips out of the tank;
5. A reservoir in which oil is heated for adding to the circulating frying oil and
6. Vapour-collecting hoods above the tank.

Temperatures normally used are from 350 to 375° F at the receiving end and 320 to 345° F at the exit end.

The oil used for deep-fat frying of potato chips has two functions:

i. It serves as a medium for transferring heat from a thermal source to the tuber slices;
ii. It becomes an ingredient of the finished product.

Use of highly refined oil is of great importance in flavour and stability of the crisps. Flavour, texture, and appearance are affected both by the amount of oil absorbed and its characteristics as it exists in the crisp (i.e. not necessarily its initial chemical and physical parametres).

Oils change continuously during the frying process but the heat abuse resulting from the crisp cooking is relatively mild. Temperatures rarely rise above 385° F at any point. Better control over crisp colour could be obtained if the final stage of moisture removal could be achieved without the browning reaction that always accompanies it in the frying process.

Crisps may be sorted for size after frying, with the larger crisps being diverted to the bulk packs and larger pouches and the smaller pieces used for vending machine packs and other individual service containers. Potato crisp sizing is also accomplished by separating the peeled potatoes into large and small sizes, which are then sliced and fried separately.

The crisps are salted immediately after they leave the fryer. It is important that the fat be liquid at this point to cause maximum adherence of the granules. Powders containing barbecue spices, cheese, or other speciality materials may be added at this point. The salt may contain added enrichment materials or antioxidants.

After salting, the crisps pass on to a conveyor belt where they are visually inspected and off-colour material is removed. I the crisps are allowed to cool before packaging, better adherence of salt and flavour powders is obtained.

Some consumers prefer the hard, curled-up crisp that is characteristic of the hand-kettle type of operation. The special flavour of the hand-kettle crisp is said to be due, at least partly, to the starch retained on the cut surfaces of the potato slices as a result of the omission of a washing process after slicing. Starch-covered slices tend to stick together in the fryer so it is necessary to use devices to prevent clumping.The principal factors affecting potato crisp acceptability are piece size, colour, and of course, flavour. These factors are controllable primarily by selection of the raw material, adjustment of processing conditions, and packaging.

STORAGE STABILITY

If the frying oil is stabilised and has not deteriorated through use, and if the packaging is opaque and has a low moisture vapour transmittance rate (MVTR), a shelf-life of 4-6 weeks should be achieved when crisps are stored at temperatures of about 70°F.

Once potato crisps are in the bag, the three forms of quality loss which have the greatest effect on consumer acceptance are breakage, absorption of moisture with loss of crispiness, and fat oxidation leading to development of rancid odours.

The mechanical abuse causing breaking of the crisps can be partially prevented by using stiff packaging material, making the package "plump" with contained air, and avoiding crushing in the shipping case. Absorption of moisture is prevented largely by proper choice of packaging material. Cellophane coated with various moisture barriers has proved to be a satisfactory pouch films for the relatively short shelf-life expected (generally stated to be 4-6 weeks).

Light (especially fluorescent light) accelerates oxidation, so that opaque packaging material must be used to obtain maximum shelf-life.

Potato crisps are considered commercially unacceptable when they have a moisture content above 3%, which is in equilibrium with a relative humidity of about 32%. The containers should have a high degree of resistance to moisture-vapour transfer. If pouches are used, foil-containing films are preferable, since they not only resist moisture-vapour transfer but reflect light.

VEGETABLE JUICES AND CONCENTRATED PRODUCTS

VEGETABLE JUICES

Vegetable juices are natural products constituted from cellular juice and a part of crushed pulp, from the tissues of some vegetables. These juices contain all valuable substances from the vegetables: vitamins, sugars, acids, mineral salts and pectic substances. The most important of these products is tomato juice; in a lower proportion there are also other juices (carrots, beet, sauerkraut, etc.).

TOMATO JUICE

This product is characterised not only by its organoleptical properties (taste, colour, flavour) but also by its vitamin content close to those of fresh tomatoes. Modem technology is oriented to a maximum maintenance of organoleptic properties and of vitamin content. At same time, it is important to assure juice uniformity by avoiding cellulosic particle sedimentation. Juice stability is assured by a flash pasteurization which assures the destruction of natural micro-flora, while keeping the initial properties.

The modern technological flow-sheet covers the following main operations: Pre-Washing is carried out by immersion in water, cold or heated up to 50° C (possibly with detergents to eliminate traces of pesticides). This operation is facilitated by bubbling compressed air in the immersion vessel/equipment.

Washing is performed with water sprays, which in modern installations have a pressure of 15 at or more. Sorting/Control on rolling sorting tables enables the removal of non-standard tomatoes - with green parts, yellow coloured, etc.

Crushing in special equipment. Preheating at 55-60° C facilitates the extraction, dissolves pectic substances and contributes to the maintaining of vitamins and natural pigments. In some modern installations, this step is carried out under vacuum at 630-680 mm Hg and in very short time.

Extraction of juice and part of pulp (maximum 80%) is performed in special equipment/tomato extractors with the care to avoid excessive air incorporation. In some installations, as an additional special care, a part of pulp is removed with continuous centrifugal separators.

De-Aeration under high vacuum of the juice brings about its boiling at 35-40° C. Homogenisation is done for mincing of pulp particles and is mandatory in order to avoid future potential product "separation" in two layers.Flash Pasteurization at 130-150° C, time = 8-12 see, is followed by cooling at 90° C, which is also the filling temperature in receptacles (cans or bottles).Aseptic Filling Closing of receptacles is followed by their inversion for about 5 to 7 minutes. Cooling has to be carried out intensely. Full cans do not need further pasteurization because the bacteria that have potentially contaminated the tomato juice during filling are easily destroyed at 90° C due to natural juice acidity. For bottles, it may be possible to avoid further sterilisation if the following conditions can be respected: washing and sterilising of receptacles, cap sterilisation (with formic acid), filling and capping under aseptic conditions, in a space with UV lamps. In so far as this is quite difficult to achieve it may be necessary to submit bottles to a pasteurization in water baths.

The main characteristics of high quality tomato juice are:

- Natural red colour;
- Taste and flavour of fresh tomatoes;
- Uniformity (without pulp sedimentation);
- Total soluble solids: 6% minimum;
- Total soluble substances (by refractometre): 5% minimum;
- Vitamin C: 15 mg/100ml minimum.

In traditional processes it is recommended to:

- Thoroughly wash and rinse the empty receptacles (including jar caps/ covers and bottle crown corks) and then "sterilize" by keeping in boiling water for 30 mini
- Add salt and lemon juice to the prepared receptacles just before filling;
- Pasteurize closed glass receptacles (bottles or jars) according to conditions recommended in technological flow-sheets and which is summarised as follows:

Receptacle size	Pre-heating	Time of pasteurization
0.33 1	60° C	40 minutes
0.501	60° C	45 minutes
0.66 1	60° C	55 minutes
0.751	60° C	60 minutes
1.0 litre	60° C	70 minutes

CARROT JUICE

This product represents an important dietetic product due to its high soluble pectin content. Technological flow-sheet is oriented to the maintaining of as high as possible a pectin content and covers the following steps:

- PRE-WASHING
- CLEANING
- WASHING
- BLANCHING in steam for 20 minutes
- GRATING
- PRESSING
- JUICE In the pressed juice will then be incorporated 25% of grated carrot (non pressed)
- HOMOGENISATION in colloidal mills
- ACIDIFICATION with 0.25% citric or tartric acid
- DE-AERATION
- FILLING in receptacles (bottles or tinplate cans)
- AIRTIGHT SEALING
- Pasteurization at 100° C for 30 minutes.

The main characteristics of a good quality carrot juice:

- Uniformity (no separation in layers occurs during storage);
- Good orange colour;
- Pleasant taste, close to fresh carrot taste;
- Total soluble solids: 12 %;
- Total sugar content: 8%;
- Beta-carotene: 1.3 mg/100 ml;
- Soluble pectin: 0.4 %.

RED BEET JUICE

The product is obtained following this technological flow-sheet: washing, cleaning, steam treatment/steaming (30-35 min at 1050 C), pressing, strain through small hole sieve, filling in receptacles, tight sealing/closing, sterilisation (25 min at 1 15° C). In order to improve taste, the juice is acidified with 0.3% citric or tartric acid.

SAUERKRAUT JUICE

Sauerkraut juice is produced in some countries for its dietetic value (lactic acid and vitamin C content) and its refreshing taste. The juice which is the result of the fermentation of lactic acid from cabbage, mainly from sliced sauerkraut, is used. The juice must be the result of a normal lactic fermentation, i.e. without butyric fermentation or other deterioration. A good quality juice must have an acidity of 1.4% lactic acid and a content of maximum 2.5% salt; this is obtained by the mixing of various sauerkraut qualities.The collected juice (from sauerkraut production) is heated slightly in order to eliminate CO_2 gas and to obtain protein coagulation. Filtration of juice is the next technological step, followed by filling in receptacles, closing of receptacles and pasteurization at 75-80° C for 4-5 minutes.

CONCENTRATED TOMATO PRODUCTS

TOMATO PASTE

The product with highest production volumes among concentrated products is tomato paste which is manufactured in a various range of concentrations, up to 44% refractometric extract. Tomato paste is the product obtained by removal of peel and seeds from tomatoes, followed by concentration of juice by evaporation under vacuum.

In some cases, in order to prolong production period, it may be advisable or possible to preserve crushed tomatoes with sulphur dioxide as described under semi-processed fruit "pulps".

Technological flow-sheets run according to equipment/installation lay-outs, which are especially designed for this finished product. Manufacturing steps fall into three successive categories:

a. Obtaining juice from raw materials;
b. Juice concentration and
c. Tomato paste pasteurization.

a. Obtaining juice from raw material. Preliminary operations (pre-washing, washing and sorting/control) are carried out in the same conditions as for manufacturing of "drinking" tomato juice described above. Next operation is removal of seeds from raw tomatoes: tomato crushing and seed separation with a centrifugal separator.

Tomato pulp is pre-heated at 55-60° C and then passed to the equipment group for sieving: pulper, refiner and superrefiner with sieves of 1.5 mm, 0.8 mm and 0.4-0.5 mm respectively in order to give the smoothest possible consistency to the tomato paste.

b. Juice is concentrated by vacuum evaporation, a technological step which in modern installations runs continuously, tomato paste from the last evaporation step being at the specified concentration.

In continuous installations with three evaporation steps (evaporating bodies), the juice is submitted in step/body I to pasteurization at 85-900 C for 15 min and this will determine the microbiological stability of finished product. Vacuum degree corresponding to this temperature is 330 mm Hg.

In evaporating bodies II and III, temperatures are around 42-46° C and vacuum at 680700 mm Hg. Juice concentration occurs gradually and continuously in the three evaporating bodies.

The advantages of continuous concentration are as follows:

- The taste, colour, flavour, "shine" and consistency of tomato paste are improved because:
  i. The real concentration is performed in evaporating bodies II and III at low temperatures (42-46° C) and
  ii. The whole concentration process time from the input of juice in body I until the output of paste from body III is of about 1 hour (for paste with 30-35% refractometric extract).
- Production capacity is raised by about 30% as compared to discontinuous installations with the same evaporation surface;
- The steam consumption is reduced by 60% because heating of bodies II and II is done with vapours resulting from juice evaporation in body I (double effect); water and electricity consumptions are also reduced by 30-40 %.

c. Tomato paste pasteurization assures the microbiological stability of the product. For this purpose, the paste coming out from concentration equipment is passed continuously and in a "forced" mode through a tubular pasteurizer from which it emerge at a temperature of 90-92°C.

Usual commercial tomato paste types are at concentrations of 24%, 28% and 32% refractometric extract. Sometimes it is possible to obtain a tomato paste with a concentration of 44% refractometric extract; for this purpose it is necessary to eliminate a part of cellulose from tomatoes, an operation performed in a separating turbine.

Tomato paste storage and preservation is carried out after packing which is done usually in drums, metallic cans or glass jars; some modern equipment has been developed for packing in aluminium bags. As far as the concentration of tomato paste is concerned it is not possible to reduce water content down to 30% which corresponds to a water activity aw of 0.700.75 (minimum limit of mould growing), it is necessary to take special measures (e. g pasteurization, cold storage or salt addition). Salt is not a preservative in itself but contributes to the lowering of water activity.

In drums, the preservation of tomato paste with minimum 30% refractometric extract is carried out in two ways:

- The hot paste (about 90°C) flows directly from pasteurization equipment into drums that have been previously steamed;
- The paste is cooled down to 30°C through a heat exchanger and is introduced into drums that have been previously steamed.

For preservation purposes, it is possible to add 3-8% salt.

Preservation with 3% salt must be carried out respecting the following criteria:

a. Processing of a healthy raw material;
b. Thorough washing and control;
c. Pasteurization of concentrated paste and use of well prepared drums. Paste in drums has to be stored in cold storage rooms during the hot season.

Preservation in big metal cans of 5 and 10 kg capacity of tomato paste with a minimum of 30% refractometric extract can be achieved without sterilisation if the following conditions are respected:

a. Sterilisation by steam of cans and covers;
b. Filling of paste at 92-94°C;
c. Airtight sealing/closing of cans;
d. Invert cans and then
e. Air cooling.

For small packages (tinplate cans of 1/10-1/1 or glass jars of same capacity) it is usual to use pasteurized paste, as hot as possible (92-94°C). The receptacles are first sterilised by steam. After airtight sealing, the receptacles are kept in boiling water for a short time in order to sterilize their inner surface and the paste in contact with inner receptacle surface.

In some countries small receptacles are not further sterilised if the manufacturing is carried out in perfect hygienic and sanitary conditions. Packing in small tinned aluminium tubes is carried out with concentrated paste, pasteurized and hot.

Good quality tomato paste is an homogenous mass, with a high density, without foreign bodies (seeds, peel, etc.), with a red colour, and an agreeable taste and smell, close to those of fresh tomatoes.

There are usually three types of tomato paste: 36, 30 and 24 which have refractometric extracts of respectively 34-38%, 28-32% and 24-26%. Paste of good quality must have a volatile acidity of maximum 0.15% as lactic acid. An 8% salt addition is accepted.

CONCENTRATED TOMATO JUICE

Concentrated tomato juice is a product with 17-19 % refractometric extract and is a homogenous mass, finely sieved, without foreign bodies/and without any evidence of deterioration. A good quality product has a red colour, an agreeable and specific taste and smell. Modern technology uses the same installations, equipment and flow-sheets for concentrated tomato juice as for the production of tomato paste; the final concentration is thus regulated between the above specified limits.

The concentrated tomato juice is filled in receptacles (metal tinplate cans or glass bottles) and then pasteurized at 100°C during 15-25 minutes according to receptacle type. With modern production lines it should be possible to pass the concentrated tomato juice through a tubular pasteurizer and then pack aseptically and cool, without the need to pasteurize the receptacles.

TOMATO SAUCES

Under the USA Code of Federal Regulation 7 CFR 52, 1991 tomato sauce is the concentrated product prepared from the liquid extract from mature, sound, whole tomatoes, the sound residue from preparing such tomatoes for canning, or the residue from partial extraction of juice, or any combination of these ingredients, to which is added salt and spices and to which may be added one or more nutritive sweetening ingredients, a vinegar or vinegars, and onion, garlic, or other vegetable flavouring ingredients.

The refractive index of the tomato sauce at 20° C is not < 1.3461. These products are widespread in some countries and are used in order to spice some meals. Sauces can be obtained from fresh tomatoes or from concentrated products (tomato paste or concentrated tomato juice), those from fresh tomatoes being of superior quality.Technological processing covers the following steps: concentrated juice processing, addition of flavour/taste ingredients (salt, sugar, vinegar, spices, etc.), boiling, fine sieving, filling of receptacles, closing and pasteurization (45 min at 85° C). Tomato sauces which can be sweet, more or less spicy are prepared according to specific recipes.

PRODUCTION ACCIDENTS AND PRODUCT DEFECTS; MEANS TO AVOID THEM

TOMATO JUICE

- "Separation" in layers is due to not enough homogenisation or low/insufficient viscosity. In the first case it is necessary to intensify homogenisation; and in second to increase the pre-heating temperature to 60° C in order to obtain protopectine hydrolisis and pectolitic enzymes inactivation.

- Moulding of the juice is brought about by significant infection of raw materials, inadequate washing and control or by use of contaminated packages. The preventive measures should be decided after cause analysis. Good pasteurization can destroy all moulds but the bad juice taste remains.
- Fermentation of juice is manifested by a significant development of gases. Prevention methods are the same as for moulding.
- Tomato juice turns sour, without the formation of gases; this defect is initiated by thermophyl and thermo-resistant bacteria; the juice acquires a vinegary taste. Prevention: maintenance of flash pasteurization temperature at 130-135° C.
- Excessive vitamin C losses are due to a simultaneous action of heating and oxygen from air. Prevention:
  a. Prevent air going into crusher and extractor;
  b. Assure an intensive de-aeration (vacuum degree 700 mm Hg) at a temperature of at least 35-40° C; and
  c. Close receptacles in vacuum.
  i. Weak colour of tomato juice can be avoided by the utilisation of mature tomatoes and with a pulp of as red a colour as possible.

TOMATO PASTE AND CONCENTRATED JUICE

- Presence of sand is caused by inadequate washing or by a significant contamination of raw material; this can be prevented by a more intensive pre-washing and washing of tomatoes.
- There may be mould especially at the surface of tomato paste packed in drums. Prevention:
  a. Accurate pre-washing and washing;
  b. Follow pasteurization instructions;
  c. Pack in clean drums or receptacles; and
  d. Close receptacles immediately after filling.
  i. Fermentation is manifested by a weak alcohol smell or by a weak vinegar taste; when the fermentation is more advanced there is gas production in the product mass. Prevention: as for moulding prevention.

TOMATO SAUCES

- Surface of the product turns black at the contact zone with air; this is due to the action of iron on the tannins from spices, tomato seeds, etc. Prevention:
  a. Avoid iron equipment;
  b. Avoid crushing of tomato seeds and
  c. Seal receptacles in vacuum.

PICKLES AND SAUERKRAUT TECHNOLOGY

VEGETABLE NATURAL ACIDIFICATION TECHNOLOGY

GHERKINS AND CUCUMBERS

Raw materials must follow strict specifications for a high quality finished product; the following parametres must be considered as critical:

- Adapt a uniform size according to the finished product requirements; for example, gherkins will need to have a maximum length of 9 cm for raw vegetables. Generally 15 cm size/length will be a maximum for high quality cucumber products in many countries. However, according to local preferences, bigger cucumbers could be also in demand.
- Cylindrical or ovoidal shape;
- Dark green colour;
- Absence of surface defects due to cryptogamic diseases.

Cucumbers have to be picked at their ripeness for eating, when the sugar content is at about 1.5-2.2%, needed for lactic fermentation. Unripe cucumber does not have enough sugar.

The general technological flow-sheet is as follows:

- RECEPTION
- CONTROL
- TEMPORARY STORAGE
- GRADING BY SIZE
- WASHING

Small holes are made in large size cucumbers skin:

- *Rreceptacle filling:* raw material is simply put in the receptacles in bulk, with care to arrange them in such a way that a maximum of pieces could be introduced;
- *Salt solution preparation:* 6% salt solution (NaCl);
- *Salt solution addition*: the salt solution is poured into the receptacle;

Fermentation is carried out at 20-30° C, anaerobically. This step takes generally 4-8 weeks. Acidity reaches a value up to 1.5% lactic acid (and in some exceptional cases up to 2% lactic acid) which corresponds to a maximum pH value of 4.1.

Storage; after the last fermentation stage, drums and other receptacles have to be stored at low temperature; best conditions for 12 months shelf life should be below + 15° C. Storage temperature will determine the shelf life of the products. Addition of 1000 ppm potassium sorbate will prevent mould development without having any influence on lactic fermentation. Raw material grading by size is a very important technological step. In order to accelerate brine penetration, mainly for medium to large size cucumbers, the practice of making small holes in the raw material skins is generally recommended. A major factor influencing the quality of lactic fermented cucumbers is the water durity; optimal results are obtained at 15-20° durity. Cucumber consistency/ texture is influenced by the formation of calcium pectate with the pectic substances from raw material tissues. In some countries, calcium chloride (0.3-0.5 %) is added in order to firm up the cucumber consistency. Chlorinated water which still contains active chlorine can inhibit or even stop the lactic fermentation.

SAUERKRAUT

In some countries cabbages are submitted to lactic fermentation as whole vegetables; however, in many countries the cabbage is shredded before fermentation. As shredded cabbage and its technology is at the basis of an important industry, giving good quality products, with a uniform fermented product and with good keeping quality and ease of distribution, this will be described first.

Cabbage as raw material for sauerkraut must be sound, ripe for eating, well-leafed and from suitable varieties. Optimum total sugar level needed for the lactic fermentation is 24%; generally good quality raw material contains up to 30-60 mg/100 g of vitamin C.

SHREDDED SAUERKRAUT

The technological flow sheet is as follows:

- RECEPTION
- CONTROL

Temporary storage is carried out in bulk, up to a height of about 1 m, during few days. This step produces a heat generation which facilitates later fermentation by the softening of tissues.

REMOVAL OF EXTERNAL LEAVES

Coring is done with a specially adapted mechanical screw; this operation generates small particles of finely divided cabbage which will be mixed with the main part of vegetable during shredding/chopping. The core represents about 10% from the whole cabbage, is rich in sugar and vitamin C, but being too high in fibre content needs to be chopped separately as described.

Shredding/Cutting of cabbage is carried out with complex specific equipment which is generally installed directly on the "top" of fermentation silos and is mobile, installed on rails and moves all along the silos. The dimension of resulting shredded cabbage is about 2-3 mm thick. The same complex equipment is designed to grind the added salt to fine particles and to distribute shredded cabbage and ground salt in an uniform manner to the fermentation silos. The usual capacity of fermentation silos is up to 30 tons, with separate compartments of 45 tons each.Salt addition is carried out by the equipment described above; the proportion of salt is 2-2.5% with respect to cabbage. This proportion must not be changed because the salt in this technology does not have a preservative role but only that to extract from cabbage the juice needed for fermentation.

It is be preferable to obtain a fairly light pressure on cabbage just after salt addition with some simple mechanical means. This is important in order to:

- Create an anaerobic medium for fermentation;
- Facilitate external diffusion of cellular juice;
- Assure a rational use of the fermentation space.

Fermentation: The maximum acidity level obtained is generally of about 1.5% lactic acid (and very rarely 2.5%); this is obtained in 4-6 weeks. Optimal acidity is 1.0-1.8% and pH value 4.1 or lower.

Fermentation temperature is at 20-25° C in the first phase and needs to be lowered then to 14-18° C. During fermentation, the brine from each storage/ fermentation silo cell is periodically circulated with a pump in order to uniformise the fermentation process.

Storage is performed in same silos used for fermentation, or the finished products is removed from silos and packed in drums and other receptacles according to distribution schedule. These silos are usually made of reinforced concrete and coated with gritstone plates or with an acid-resisting material layer.

At small scale and in traditional processing, shredded sauerkraut can be obtained by using simple available glass or rigid plastic receptacles. At home, this process can use glass jars and/or local/traditional pottery receptacles from a minimum size of 2-3 kg up to the available/practical sizes (better limited to 10-15 kg).In some countries shredded sauerkraut is preserved in receptacles by pasteurization, once the fermentation process has been completed.

WHOLE SAUERKRAUT

According to the consumer preference in different countries and to the specific situations it is also usual to preserve whole cabbages by lactic fermentation. At small or medium scale operations, whole cabbage could be processed/fermented in cylindric receptacles like 30 to 200 litre rigid plastic drums, or rectangular receptacles made from food grade rigid plastic. It is possible to find this type of drum in a significant number of developing countries. These two types of rigid plastic receptacles could also be used for shredded sauerkraut production.

Prepared whole cabbages are put into fermentation receptacles and a 5-6 % salt concentration brine is poured on top. The fermentation conditions are the same as for shredded sauerkraut. In order to assure a uniform fermentation and to avoid a strict anaerobic (butyric) fermentation it is necessary to apply a periodic juice "aeration" (each 2-3 days at the beginning of the fermentation, and then each 5-7 days).

OTHER ACIDIFIED VEGETABLES

In principle all vegetables with a sugar content of at least 2 % could be preserved by lactic fermentation. From a practical point of view it is mainly the following vegetables which are preserved by this technology: unripe tomatoes (green tomatoes), peppers, eggplant, carrots and cauliflower, alone or usually in a mix with cucumber as mixed pickles.Fermentation of individual vegetables is carried out according to a flow-sheet as described for whole sauerkraut. The type of cut, brine concentration and frequency of operating steps have to be adapted to each case; green tomatoes are fermented as whole vegetable.

SIMPLIFIED FLOW-SHEET FOR WHOLE SAUERKRAUT PROCESSING

Process

a. Cabbage preparation
   - Remove the damaged leaves;
   - Wash the vegetable;
   - Remove 2-3 outer leaves;
   - Size grade in three categories:

   i. Size A: about 700 g per cabbage

   ii. Size B: less than 1.2 kg per cabbage

   iii. Size C: more than 1.2 kg per cabbage
   - Process each size category separately;
   - Wash cabbages;
   - Remove cores;
   - Cut size category C vegetable in halves.

b. Salt solution (brine) preparation
   - Prepare a 5% salt (NaCl) solution = 500 g salt for 10 litre water or 50 g salt for 1 litre water;
   - Stir until complete salt dissolution;
   - Filter salt solution through cheese cloth.

c. Initial processing
   - Use a different receptacle for each size category;
   - Arrange cabbages in fermentation receptacle;
   - Pour salt solution to completely cover cabbages;
   - Fix some clean wood pieces (or better some fitted covers with holes) in order to keep cabbages completely covered by salt solution. Allow about 10 cm salt solution above cabbage level;

- Store fermentation receptacles in a moderately cold and ventilated place, out of direct sunlight/heat, protected from dust and other nuisances (insect, etc.);
- Cover each fermentation receptacle with a piece of cardboard or cloth.

d. Processing follow-up
- During the first week after initial processing. Once every 2 days, it is necessary to: remove the cover;
- Collect and carefully remove carefully (with a household spoon) the white layer ("scum") formed at the surface of salt solution;
- Wash the spoon each time and rinse;
- Put back the cover.
- During the 5 following weeks.
- Once every 4 days: repeat the operations described above.

After the first week, in order to assure a homogeneous acidification/ fermentation process for big receptacles (i.e. drums or other receptacles of 20 to 2001 capacity), it will be necessary to proceed once a week to an "aeration" step.

After completion of brine surface cleaning (as described above), the following operations will be carried out:

- Remove the cover and the wood spacers;
- Remove all salt solution (brine) from the receptacle;
- Filter this solution through a cheese cloth;
- Pour back the filtered solution back to the fermentation receptacle;
- Put the wood spacers back in place;
- Cover the receptacle.

These operations will be carried out for each fermentation receptacle once a week, during an estimated period of six weeks; total duration will be determined by the temperature in the storage room and by the chemical composition of specific raw material (cabbage) lots. Always keep salt solution (brine) level at 10 cm above cabbages, e. g. cabbages must be always covered by brine.

CONSUMPTION OF THE FINISHED PRODUCT

It is possible to estimate that at reasonable ambient temperatures and with a strict follow up of the above recommendations, the finished product will be ready for consumption about 6 weeks after initial processing.

The finished product could be used "as is" in vegetable salads, or prepared according to local taste: with tomato sauce, beans, minced meat, etc. as a replacement of fresh cabbages. In the same way as with natural acidification or lactic fermentation the cabbage texture is modified and softened so that tissues are more digestible than fresh vegetable. It is possible to use the finished product in local dishes and in new recipes without having to boil it. Apart from the taste benefits of acidified cabbages, this is also produces a significant fuel savings.The juice resulting from natural cabbage acidification is recovered and could be used separately as a refreshing vegetable juice; the preparation is described in this document.

FINISHED PRODUCT STORAGE

It is possible to store the finished product after completion of fermentation (i.e. after the estimated six weeks period); the storage time will depend on the ambient air temperature. If a cool space is available, the finished product shelf-life/storage time at a temperature of about + 15° C is estimated at six months. At an ambient temperature not exceeding +20° C, the storage time could be estimated at 2-3 months.

ARTIFICIAL VEGETABLE ACIDIFICATION TECHNOLOGY

This technology is based on the addition of food grade vinegar which has a bacteriostatic action in concentrations up to 4 % acetic acid and bactericidal action in higher concentrations.

Vegetables preserved in vinegar need to reach, after equilibrium between vinegar and water contained in vegetables, a final concentration of 2-3 % acetic acid in order to assure their preservation.

To achieve this final concentration, a 6-9 % acetic acid vinegar is used, as related to the specific ratios vinegar/vegetables. In vinegar pickles, salt (2-3 %) and sometimes sugar (2-5 %) are also added.

If the vinegar concentration is lower than 2%, vinegar pickles need to be submitted to a pasteurization in order to assure their preservation.

CUCUMBERS IN VINEGAR

This represents the basic product obtained by this technology. Cucumbers have to be wholesome, with a soft texture and not have reached eating maturity. They must have a low sugar content because in this technology there is no lactic fermentation involved. Dimensions are up to 12 cm length, with a preference for small cucumbers.

The technological steps are the followings:

- SIZE GRADING
- WASHING
- ARRANGE IN RECEPTACLES - glass jars, etc.

Pouring of vinegar is usually carried out at room temperature; however, hot vinegar addition enables a sterilisation of cucumber surface and facilitates vinegar penetration in vegetable tissues.

- SALT (SUGAR) ADDITION
- SPICING ADDITION

The technological cycle of artificial acidification is considered completed when acetic acid concentration reaches an equilibrium value; the time needed is about 2 weeks.

When equilibrium concentration in acetic acid is below 2 %, the cucumbers are submitted to a pasteurization for 20 min at 90-1000 C in order to assure their preservation.

Cucumbers in vinegar with previous lactic fermentation are excellent quality products because the lactic fermentation improves the taste of these cucumbers. The principle of this process is to assure preservation both by acetic acid and by lactic acid simultaneously.

Technological processing flow-sheet is as follows: small cucumbers ("cornichons" or "gherkins") are washed, brushed and small holes are made in the skin; the vegetables then are put in drums with slightly warm 6% brine which also contains spices.

The lactic fermentation runs for few days up to a lactic acid concentration of about 0.5 %. The cucumbers are removed from the brine, washed thoroughly and well drained. Preservation is usually done in glass jars by pouring a normally flavoured vinegar with about 9% acetic acid usually in order to bring the final concentration to 3% calculated as acetic acid.

In order to obtain a high quality product only wine vinegar should be used. In some pickles (e.g. in "Cornichons") the usual level of wine vinegar is set at 20 % of packaged product total weight; some alcohol vinegar could be still added and final concentration will be adjusted as described above.

VEGETABLE CANNING

Canned vegetables can be classified as follows:

1. Canned products in salt brine;
2. Canned products in tomato concentrated juice;
3. Canned products in vegetable oil.

CANNED VEGETABLES IN VEGETABLE OIL

General flow-sheet is described below:

- Reception
- Sorting
- Cleaning/peeling
- Washing
- Cutting
- Frying or Blanching
- Cooling
- Filling and adding of vegetable oil, sauce or tomato concentrated juice
- Sealing
- Sterilisation

- Cooling
- Labelling
- Storage

GENERAL HEAT PRESERVATION OPERATIONS – CANNING

The success of heat preservation operations lies in:

- Selecting suitable fruit and vegetables in good conditions;
- Preparing them hygienically and skilfully;
- Packing them in cans which are hermetically sealed and then processed under fixed conditions of time and temperature;
- Cooling these cans carefully and storing them under conditions which will not cause deterioration of either the cans or their contents.

SELECTION OF RAW MATERIALS

It is appreciated that some varieties of fruit and vegetables are not suitable for canning, either because they are uneconomical to prepare or because the colour, flavour or texture are poor.Suitable varieties must be available to the canner in quantities sufficient to meet his requirements and in sound conditions for canning. The flow to the cannery should be regulated in order that perishable materials are not left for a long time before being handled, since any delay will cause deterioration.Apart from the main ingredients, be it fruit or vegetables, minor ingredients also require careful selection. Sugar, salt, water and spices for instance may all be contaminated with spoilage organisms, so constant testing of all raw materials is essential.

PREPARATION

This is carried out by various methods, including grading, trimming, peeling, washing and blanching. All equipment must be scrupulously clean and preparation should be completed quickly and carefully in order to keep the bacterial load as low as possible. Thorough washing of vegetable is necessary to remove spores of heat resistant bacteria which are present in large numbers in the soil.

Blanching in steam or hot water is of no avail against these heat resistant (thermophilic) spores because of the comparatively low temperatures involved.

Reasons for blanching are:

- The removal of gas from the tissues of the raw material;
- The shrinkage of this material;
- The inhibition of enzymic reactions, which, if not checked, will adversely affect the colour and nutritive value of the food.

FILLING

Filling, be it mechanical or by hand, requires careful attention. The cans must be clean and the correct weight of foodstuffs must be added. Under-filled cans will be underweight and the headspace will be too large, resulting in too much air being left in the can. Overfilling may lead to seams being strained during processing and to ends becoming distorted and bulged.

If the product forms hydrogen on storage as is the case with coloured fruits, swelling of the can due to hydrogen pressure will occur more quickly in an overfilled can than in one which has been correctly filled. Overfilling also affects heat penetration in the can and may lead to spoilage outbreaks.

AIR REMOVAL

Before the can is seamed, air must be removed from the contents and the headspace. Normally, this is carried out by passing the cans through a steam box until the temperature at the centre of the can is at least 160° F.

This operation, termed exhausting, is necessary for the following reasons:

i. To minimise strains on the seams due to expansion of air during the processing period;
ii. To remove oxygen which accelerates corrosion in the can and also causes oxidation of the food with possible serious effects on colour and flavour;
iii. To reduce the destruction of vitamin C;
iv. To enable a vacuum to be formed when the can is cooled.

This ensures that the ends remain concave, even when storage temperatures are a little higher than usual, and also acts as a reservoir for hydrogen which may be formed by reactions between the can and its contents. Thus a high vacuum makes for a long shelf life. Large cans, however, should not reach such a high exhaust temperature before seaming as smaller cans because of the danger of the can body collapsing on cooling, a condition known as "panelling".

DOUBLE SEAMING

The can should be double-seamed as soon as the correct centre temperature has been attained. Any delay between exhausting and seaming will lead to loss of vacuum and may lead to bacterial spoilage. The quality of the double seam must, of curse, be frequently checked.

HEAT PROCESSING

After seaming, the cans are heated for a definite time at a definite temperature to kill or inhibit organisms which may cause spoilage. This operation is termed "heat processing". The times and temperatures required for "heat processing" of various packs have been determined experimentally to ensure that spores of the most heat resistant food poisoning organisms known, Clostridium botulinum, are destroyed. There are other organisms, however, whose spores are more heat resistant than those of Clostridium botulinum and which although they will not cause food poisoning may cause spoilage and for this reason the minimum heat processing time is often exceeded by recommendations made by laboratories. At the same time there is a limit to the amount of heating which a canned food may be given without spoiling its flavour, texture and colour and this also has to be taken into consideration when process recommendations are made.

Bacterial spores have a greater resistance to heat when the growth-medium is neutral or near neutral, and neutrality is normally required for bacterial growth to commence.

Because of this, canned foods have been broadly divided into two groups:

a. "acid" foods having a pH of 4.5 or lower and

b. "non-acid" foods having a pH of more than 4.5.

"Non-acid" foods (vegetables) must, therefore be "heat processed" at high temperatures using steam under pressure, whereas "acid" foods (fruit) may be processed at the (lower) temperature of boiling water, since this will kill moulds and yeasts and if any bacterial spores survive the combination of acid and heat, they will be inhibited from growth by the acid environment.

The rate of destruction by heat follows a definite pattern, the same proportion of the surviving bacteria being destroyed in successive units of time. The more bacteria there are in a pack, the more time will be need to reduce their numbers. For this reason, it is essential that the initial number of bacteria be kept low, and this may be achieved by ensuring fast and hygienic handling at all stages in the cannery.

Pressure gauges and retort temperature control equipment must be checked frequently for accuracy. Processing times and temperatures must be strictly adhered to, and complete removal of air from the retort during processing must be achieved by adequate venting. Failure to remove the air completely will result in their being cold spots in the retort and intermittent spoilage are likely.

COOLING

As soon as the heat processing time is completed, the cans are cooled in chlorinated water as rapidly as possible without damaging them. Cans processed in steam develop high internal pressure because of the expansion of the foodstuff, the expansion of air in the can and the increase in the vapour pressure of the water in the can.

During the heat process, these pressures are counter-balanced to some extent by the pressure of the steam in the retort, but on releasing this steam pressure at the commencement of the cooling period, the pressure in the can may be sufficient to strain the seams seriously and may even distort the ends.

Cans of A21/2 size or larger, when processed at temperatures of 240° F or more, are liable to undergo permanent distortion, such as peaking. This may be avoided by pressure cooling, which involves replacing steam pressure by air pressure before introducing water to the retort, and maintaining this until the pressure inside the can has fallen to a safe level.

This presents difficulties, since if the air pressure is maintained after the can has developed a vacuum, the can body is liable to collapse. Where pressure-cooling is not carried out, the retort pressure is allowed to drop slowly to atmospheric pressure and the cans are then cooled with water.

STORAGE

After cooling, the cans should be stored in cool, dry conditions. The maintenance of a constant temperature is desirable, since a rise in temperature may lead to condensation of moisture on the can, with possible rusting. Cool conditions are required because storage at higher temperatures not only causes chemical and physical changes in the product and the container but also introduces a risk of thermophilic spoilage.

Other known causes of container spoilage in storage are the use of labels and cardboard cases which have too high a chloride content, and the use of unseasoned wood in the manufacture of packing cases, all of which tend to cause rust formation on the cans.

General technical operations for fruit and vegetable canning lines:

a. Receptacle washing will remove the impurities and, as much as possible, the microorganisms on the inner surface of metallic cans or glass jars. Washing must be performed just before receptacle filling in order to avoid a new contamination.

   Washing methods are variable and depend on receptacle type and need to be carried out with adequate mechanical equipment. Metal cans are washed on the can feeding lines of filling equipment; a high pressure spray of warm water (65-8° C) is directed into the receptacles while these are submitted simultaneously to a rotation and forward motion.

   Glass jars are submitted to a triple washing: wetting for 10 min in a warm detergent and disinfectant solution (40-45° C) containing 100 mg active Cl/litre; washing with high pressure (2.5-3 at) warm water sprays (65-85° C); rinse with cold water. Special attention MUST be given to recycled glass jars; washing process must be intensified or repeated, depending upon their contamination.

b. Receptacles are filled in order to maintain a specific ratio between the solid part of the composition and the filling or covering liquid.

   For canned vegetable products, the covering liquid may be a 1-3% salt solution with or without addition of sugar (1-3%), tomato concentrated juice or various sauces based on concentrated tomato juices. Salt solution (brine) preparation may be performed with salt percolators; the resulting solution is saturated, containing 318 g/l and needs to be diluted to usual concentrations (1-3%). Brine is then heated up to filling temperatures which depend on product type (up to 85-90° C).

Sugar solutions (syrups) for fruit products may be prepared on the same type of percolators as brine. Receptacle filling is carried out by leaving an empty space of 5-15% of the total volume, depending on filling temperature and the product type.

c. Pre-heating (exhausting) of full receptacles aims at the removal of air from the tissues and the increase of the initial temperature of the receptacle contents. On modern production lines, exhausting is eliminated and replaced by the increase of the filling liquid temperature and hermetic receptacle closing under vacuum.

When exhausting is applied, with steam or with hot water, the pre-heated receptacles must be immediately closed in order to avoid the contraction of liquid phase and thus air introduction. Exhausting is performed in special, continuous equipment; product temperature is between 80 and 95° C, during 2-10 min.

5

Methods of Reducing Deterioration

A knowledge of deterioration factors and the way they act, including the rates of deterioration to a specific category of food, means that it is possible to list the ways of lowering or stopping the action and obtaining fruit and vegetable preservation. In order to maintain their nutritional value and organoleptic properties and because of technical-economical considerations, not all the identified means against deterioration actually have practical applications for fruit and vegetable preservation.

TECHNICAL METHODS OF REDUCING FOOD DETERIORATION

These technical means can be summarised as follows:

Physical	Heating
	Cooling
	Lowering of water content Drying/dehydration.
Concentration	
	Sterilising filtration
	Irradiation
	Other physical means (high pressure, vacuum, inert gases)
Chemical	Salting
	Smoking
	Sugar addition
	Artificial acidification
	Ethyl alcohol addition
	Antiseptic substance action
Biochemical	Lactic fermentation (natural acidification)
	Alcoholic fermentation

This classification of methods of reducing deterioration presents some difficulties because their preservation effects are physical, physico-chemical, chemical and biochemical complex phenomena which rarely act in isolation. Normally they take place together or one after the other.From the whole list of possible methods of reducing deterioration, over the years, some procedures for fruit and vegetable preservation have found practical application.

Procedures for fruit and vegetable preservation

Procedures	Practical applications

Fresh storage	Fruits, vegetables
Cold storage	Fruits, vegetables
Freezing	Fruits, vegetables
Drying/dehydration	Fruits, vegetables
Concentration	Fruit and vegetable juices
Chemical preservation	Fruit semi-processed
Preservation with sugar	Fruit products/preserves
Pasteurization	Fruit and vegetable juices
Sterilisation	Fruits, vegetables
Sterilising filtration	Fruit juices
Irradiation	Fruits, vegetables

These preservation procedures have two main characteristics as far as being applied to all food products is concerned:

- Some of them are applied only to one or some categories of foods; others can be used across the board and thus a wider application (cold storage, freezing, drying/dehydration, sterilisation, etc.);
- Some guarantee food preservation on their own while others require combination with other procedures, either as principal or as auxiliary processes in order to assure preservation (for example smoking has to be preceded by salting).

Combined Preservation Procedures

In practice preservation procedures aim at avoiding microbiological and biochemical deterioration which are the principal forms of deterioration. Even with all recent progress achieved in this field, no single one of these technological procedures applied alone can be considered wholly satisfactory from a microbiological, physico-chemical and organoleptic point of view, even if to a great extent the food value is assured.

Thus, heat sterilisation cannot be applied in order to destroy all micro-organisms present in foods without inducing non desirable modifications. Preservation by dehydration/drying assures microbiological stability but has the drawback of undesirable modifications that appear during storage: vitamin losses, oxidation phenomena, etc.Starting with these considerations, the actual tendency in food preservation is to study the application of combined preservation procedures, aiming at the realisation of maximum efficiency from a microbiological and biological point of view, with reduction to a minimum of organoleptical degradation and decrease in food value.

The principles of combined preservation procedures are:

- Avoid or reduce secondary (undesirable) effects in efficient procedures for microbiological preservation;
- Avoid qualitative degradation appearing during storage of products preserved by efficient procedures from a microbiological point of view;
- Increase microbiological efficiency of preservation procedures by supplementary means;

- Combine preservation procedures in order to obtain maximum efficiency from a microbiological point of view, by specific action on various types of micro-organisms present;
- Establish combined factors that act simultaneously on bacterial cells.

Research and applications in this direction were followed by microbiological and biochemical way, obtaining a serial of combination of preservation procedures with the possibility of application in industrial practice.

Fresh fruit and vegetable storage can be combined with:

- Storage in controlled atmosphere where carbon dioxide and oxygen levels are monitored, increasing concentration of CO2 and lowering that of oxygen according to fruit species. Excellent results were obtained for pomace fruit; in particular the storage period for apples has been extended. Application of this combined procedure requires airtight storage rooms.
- Storage in an environment containing ethylene oxide; this accelerates ripening in some fruit: tomatoes, bananas, mangoes, etc.

Cold storage can be combined with storage in an environment with added of carbon dioxide, sulphur dioxide, etc. according to the nature of product to be preserved.Preservation by drying/dehydration can be combined with:- freezing: fresh fruit and vegetables are dehydrated up to the point where their weight is reduced by 50% and then they are preserved by freezing.

This procedure (freeze-drying) combines the advantages of drying (reduction of volume and weight) with those of freezing (maintaining vitamins and to a large extent organoleptic properties).

A significant advantage of this process is the short drying time in so far as it is not necessary to go beyond the inflexion point of the drying curve. The finished products after defreezing and rehydration/reconstitution are of a better quality compared with products obtained by dehydration alone.

- Cold storage of dried/dehydrated vegetables in order to maintain vitamin C; storage temperature can be varied with storage time and can be at -8° C for a storage time of more than one year, with a relative humidity of 70-75 %.
- Packaging under vacuum or in inert gases in order to avoid action of atmospheric oxygen;, mainly for products containing beta-carotene.
- Chemical preservation: a process used intensively for prunes and which has commercial applications is to rehydrate the dried product up to 35 % using a bath containing hot 2 % potassium sorbate solution. Another possible application of this combined procedure is the initial dehydration up to 35% moisture followed by immersion in same bath as explained above; this has the advantage of reducing drying time and producing minimum qualitative degradation. Both applications suppress the dehydrated products reconstitution (rehydration) step before consumption.

- Packaging in the presence of desiccants (calcium oxide, anhydrous calcium chloride, etc.) in order to reduce water vapour content in the package, especially for powdered products.

Preservation by concentration, carried out by evaporation, is combined with cold storage during warm season for tomato paste (when water content cannot be reduced under the limit needed to inhibit moulds and yeasts, e.g. a_w = 0.70...0.75).

Chemical preservation is combined with:

- Acidification of food medium (lowering pH);
- Using combined chemical preservatives.

Preservation by lactic fermentation (natural acidification) can be combined with cold storage for pickles in order to prolong storage time or shelf-life.Preservation with sugar is combined with pasteurization for some preserves having a sugar content below 65%.

FRESH STORAGE

Fresh Fruit and Vegetable Storage

Once fruit is harvested, any natural resistance to the action of spoiling micro-organisms is lost. Changes in enzymatic systems of the fruit also occur on harvest which may also accelerate the activity of spoilage organisms.

Means that are commonly used to prevent spoilage of fruits must include:

- Care to prevent cutting or bruising of the fruit during picking or handling;
- Refrigeration to minimise growth of micro-organisms and reduce enzyme activity;
- Packaging or storage to control respiration rate and ripening;
- Use of preservatives to kill micro-organisms on the fruit.

A principal economic loss occurring during transportation and/or storage of produce such as fresh fruit is the degradation which occurs between the field and the ultimate destination due to the effect of respiration. Methods to reduce such degradation are as follows:

- Refrigerate the produce to reduce the rate of respiration;
- Vacuum cooling;
- Reduce the oxygen content of the environment in which the produce is kept to a value not above 5% of the atmosphere but above the value at which anaerobic respiration would begin. When the oxygen concentration is reduced within 60 minutes the deterioration is in practice negligible.

The following is a summary of some recent developments in post-harvest technology of fresh fruits and vegetables.

Harvest Maturity

This is particularly important with fruit for export. One recent innovation is the measurement of resonant frequency of the fruit which should enable the grading out of over mature and under-mature fruit before they are packed for export.

Harvest Method

Considerable research is continuing on mechanical harvesting of perishable crops with a view to minimising damage. In fruit trees, controlling their height by use of dwarfing rootstocks, pruning and growth regulating chemicals will lead to easier, cheaper more accurate harvesting.

Handling Systems

Field packing of various vegetables for export has been carried out for many years. In the last decade or so this has been applied, in selected cases, to a few tropical fruit types. Where this system can be practiced it has considerable economic advantages in saving the cost of building, labour and equipment and can result in lower levels of damage into crops.

Pre-cooling

Little innovation has occurred in crop pre-cooling over the last decade. However high velocity, high humidity forced air systems have continued to be developed and refined. These are suitable for all types of produce and are relatively simple to build and operate and, while not providing the speed of cooling of a vacuum or hydrocooler, have the flexibility to be used with almost all crops.

Chemicals

There is a very strong health lobby whose objective is to reduce the use of chemicals in agriculture and particularly during the post harvest period. Every year sees the prohibition of the use of commonly used post-harvest chemicals. New ways need to be developed to control post-harvest diseases, pest and sprouting.

Coatings

Slowing down the metabolism of fruit and vegetables by coating them with a material which affects their gaseous exchange is being tested and used commercially on a number of products.

Controlled Environment Transport

Recent innovations in this technique have produced great progress as a result of the development and miniaturisation of equipment to measure carbon dioxide and oxygen. Several companies now offer containers where the levels of these two gases can be controlled very precisely.

Water and Water Activity (a_w) in Foods

Micro-organisms in a healthy growing state may contain in excess of 80%

water. They get this water from the food in which they grow. If the water is removed from the food it also will transfer out of the bacterial cell and multiplication will stop. Partial drying will be less effective than total drying, though for some micro-organisms partial drying may be quite sufficient to arrest bacterial growth and multiplication.Bacteria and yeasts generally require more moisture than moulds, and so moulds often will be found growing on semi-dry foods where bacteria and yeasts find conditions unfavourable; example are moulds growing on partially dried fruits.Slight differences in relative humidity in the environment in which the food is kept or in the food package can make great differences in the rate of micro-organism multiplication. Since micro-organisms can live in one part of a food that may differ in moisture and other physical and chemical conditions from the food just millimetres away, we must be concerned with conditions in the "microenvironment" of the micro-organisms. Thus it is common to refer to water conditions in terms of specific activity.The term "water activity" is related to relative humidity. Relative humidity is defined as the ratio of the partial pressure of water vapour in the air to the vapour pressure of pure water at the same temperature. Relative humidity refers to the atmosphere surrounding a material or solution.Water activity or aw is a property of solutions and is the ratio of vapour pressure of the solution compared with the vapour pressure of pure water at the same temperature. Under equilibrium conditions water activity equals:

$$a_w = RH/100$$

When we speak of moisture requirements of micro-organisms we really mean water activity in their immediate environment, whether this be in solution, in a particle of food or at a surface in contact with the atmosphere.At the usual temperatures permitting microbial growth, most bacteria require a water activity in the range of about 0.90 to 1.00.Some yeasts and moulds grow slowly at a water activity down to as low as about 0.65.Qualitatively, water activity is a measure of unbound, free water in a system available to support biological and chemical reactions. Water activity, not absolute water content, is what bacteria, enzymes and chemical reactants encounter and is affected by at the micro-environmental level in food materials.

Two foods with the same water content can have very different a_w values depending upon the degree to which water is free or otherwise bound to food constituents. Fig. 5.2.1 is a representative water absorption isotherm for a given food at a given temperature. It shows the final moisture content the food will have when it reaches moisture equilibrium with atmospheres of different relative humidities.Thus, this food, at the temperature for which this absorption isotherm was established, will ultimately attain a moisture content of 20% at 75% RH (relative humidity). If this food was previously dehydrated to below 20% moisture and placed in an atmosphere of 75% RH, it would absorb moisture until it reached 20%. Conversely, if this food was moistened to greater than

20% water and then placed at 75% RH, it would lose moisture until it reached the equilibrium value of 20%.Under such conditions some foods may reach moisture equilibrium in the very short time of a few hours, others may require days or even weeks. When a food is in moisture equilibrium with its environment, then the a_w of the food will be quantitatively equal to the RH divided by 100.Qualitatively, water activity is a measure of free or available water, to be distinguished from unavailable or bound water. These states of water also bear a relationship to the characteristic sigmoid shapes of water absorption isotherm curves of various foods.Thus, according to theory, most of the water corresponding to the portion of the curve below its first inflection point is believed to be tightly bound water, often referred to as an adsorbed mono-molecular layer of water. Moisture corresponding to the region above this point and up to the curve's second inflection point is thought to exist largely as multi-molecular layers of water less tightly held to food constituent surfaces.

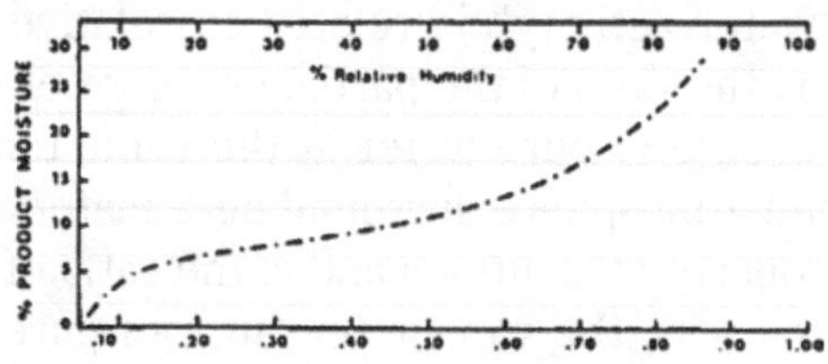

Figure: Water sorption isotherm

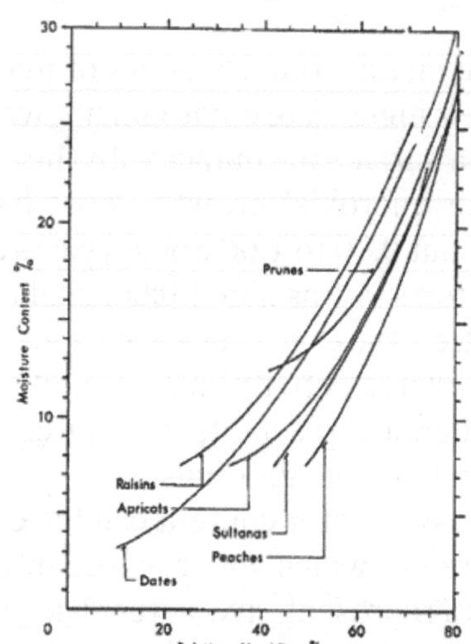

Figure: Moisture sorption isotherms for various dried fruits at 25°C

Beyond this second inflection point moisture generally is considered to be largely free water condensed in capillaries and interstices within the food. In this latter portion of the sorption isotherm curve small changes in moisture content result in great changes in a food's a_w. In Fig. given below are illustrated the moisture sorption isotherms for various dried fruits at 25°C.

Preservation by Drying/Dehydration

The technique of drying is probably the oldest method of food preservation practiced by mankind. The removal of moisture prevents the growth and reproduction of micro-organisms causing decay and minimises many of the moisture mediated deterioration reactions.

It brings about substantial reduction in weight and volume minimising packing, storage and transportation costs and enable storability of the product under ambient temperatures, features especially important for developing countries. The sharp rise in energy costs has promoted a dramatic upsurge in interest in drying world-wide over the last decade.

Heat and Mass Transfer

Dehydration involves the application of heat to vapourise water and some means of removing water vapour after its separation from the fruit/vegetable tissues. Hence it is a combined/simultaneous (heat and mass) transfer operation for which energy must be supplied.

A current of air is the most common medium for transferring heat to a drying tissue and convection is mainly involved.

The two important aspects of mass transfer are:

- The transfer of water to the surface of material being dried and
- The removal of water vapour from the surface.

In order to assure products of high quality at a reasonable cost, dehydration must occur fairly rapidly. Four main factors affect the rate and total drying time:

- The properties of the products, especially particle size and geometry;
- The geometrical arrangement of the products in relation to heat transfer medium (drying air);
- The physical properties of drying medium/environment;
- The characteristics of the drying equipment.

It is generally observed with many products that the initial rate of drying is constant and then decreases, sometimes at two different rates. The drying curve is divided into the constant rate period and the falling rate period.

Surface Area

Generally the fruit and vegetables to be dehydrated are cut into small pieces or thin layers to speed heat and mass transfer. Subdivision speeds drying for two reasons:

- Large surface areas provide more surface in contact with the heating medium (air) and more surface from which moisture can escape;
- Smaller particles or thinner layers reduce the distance heat must travel to the centre of the food and reduce the distance through which moisture in the centre of the food must travel to reach the surface and escape.

Temperature

The greater the temperature difference between the heating medium and the food the greater will be the rate of heat transfer into the food, which provides the driving force for moisture removal. When the heating medium is air, temperature plays a second important role.

As water is driven from the food in the form of water vapour it must be carried away, or else the moisture will create a saturated atmosphere at the food's surface which will slow down the rate of subsequent water removal. The hotter the air the more moisture it will hold before becoming saturated.

Thus, high temperature air in the vicinity of the dehydrating food will take up the moisture being driven from the food to a greater extent than will cooler air. Obviously, a greater volume of air also can take up more moisture than a lesser volume of air.

Air Velocity

Not only will heated air take up more moisture than cool air, but air in motion will be still more effective. Air in motion, that is, high velocity air, in addition to taking up moisture will sweep it away from the drying food's surface, preventing the moisture from creating a saturated atmosphere which would slow down subsequent moisture removal. This is why clothes dry more rapidly on a windy day.Some other phenomena influence the drying process and a few elements are summarised below.

Dryness of Air

When air is the drying medium of food, the drier the air the more rapid is the rate of drying. Dry air is capable of absorbing and holding moisture. Moist air is closer to saturation and so can absorb and hold less additional moisture than if it were dry. But the dryness of the air also determines how low a moisture content the food product can be dried to.

Atmospheric Pressure and Vacuum

If food is placed in a heated vacuum chamber the moisture can be removed from the food at a lower temperature than without a vacuum. Alternatively, for a given temperature, with or without vacuum, the rate of water removal from the food will be greater in the vacuum. Lower drying temperatures and shorter drying times are especially important in the case of heat-sensitive foods.

Evaporation and Temperature

As water evaporates from a surface it cools the surface. The cooling is largely the result of absorption by the water of the latent heat of phase change from liquid to gas.In doing this the heat is taken from the drying air or the heating surface and from the hot food, and so the food piece or droplet is cooled.

Time and Temperature

Since all important methods of food dehydration employ heat, and food constituents are sensitive to heat, compromises must be made between maximum possible drying rate and maintenance of food quality.

As is the case in the use of heat for pasteurization and sterilisation, with few exceptions drying processes which employ high temperatures for short times do less damage to food than drying processes employing lower temperatures for longer times.

Thus, vegetable pieces dried in a properly designed oven in four hours would retain greater quality than the same products sun dried over two days.

Several drying processes will achieve dehydration in a matter of minutes or even less if the food is sufficiently subdivided.

Drying Techniques

Several types of dryers and drying methods, each better suited for a particular situation, are commercially used to remove moisture from a wide variety of food products including fruit and vegetables.

While sun drying of fruit crops is still practiced for certain fruit such as prunes, figs, apricots, grapes and dates, atmospheric dehydration processes are used for apples, prunes, and several vegetables; continuous processes as tunnel, belt trough, fluidised bed and foam-mat drying are mainly used for vegetables.

Spray drying is suitable for fruit juice concentrates and vacuum dehydration processes are useful for low moisture/high sugar fruits like peaches, pears and apricots.

Factors on which the selection of a particular dryer/drying method depends include:

- Form of raw material and its properties;
- Desired physical form and characteristics of dried product;
- Necessary operating conditions;
- Operation costs.

There are three basic types of drying process:

- Sun drying and solar drying;
- Atmospheric drying including batch (kiln, tower and cabinet dryers) and continuous (tunnel, belt, belt-trough, fluidised bed, explosion puff, foam-mat, spray, drum and microwave);
- Sub-atmospheric dehydration (vacuum shelf/belt/drum and freeze dryers).

The scope has been expanded to include use of low temperature, low energy process like osmotic dehydration. As far dryers are concerned, one useful division of dryer types separates them into air convection dryers, drum or roller dryers, and vacuum dryers. Using this breakdown, Table given below indicates the applicability of the more common dryer types to liquid and solid type foods.

Table: Common dryer types used for liquid and solid foods

Dryer type	Usual food type
Air convection dryers	
kiln	pieces
cabinet, tray or pan	pieces, purées, liquids
tunnel	pieces
continuous conveyor belt	purées, liquids
belt trough	pieces
air lift	small pieces, granules
fluidized bed	small pieces, granules
spray	liquid, purées
Drum or roller dryers	
atmospheric	purées, liquids
vacuum	purées, liquids
Vacuum dryers	
vacuum shelf	pieces, purées, liquids
vacuum belt	purées, liquids
freeze dryers	pieces, liquids

Fruit and Vegetable Natural Drying — Sun and Solar Drying

Surplus production and specifically grown crops may be preserved by natural drying for use until the next crop can be grown and harvested. Natural dried products can also be transported cheaply for distribution to areas where there are permanent shortages of fruit and vegetables.

The methods of producing sun and solar dried fruit and vegetables described here are simple to carry out and inexpensive. They can be easily employed by grower, farmer, cooperative, etc.

The best time to preserve fruits and vegetables is when there is a surplus of the product and when it is difficult to transport fresh materials to other markets. This is especially true for crops which are very easily damaged in transport and which stay in good condition for a very short time. Preservation extends the storage (shelf) life of perishable foods so that they can be available throughout the year despite their short harvesting season.

Sun and solar drying of fruits and vegetables is a cheap method of preservation because it uses the natural resource/source of heat: sunlight. This method can be used on a commercial scale as well at the village level provided that the climate is hot, relatively dry and free of rainfall during and immediately after the normal harvesting period. The fresh crop should be of good quality and as ripe (mature) as it would need to be if it was going to be used fresh. Poor quality produce cannot be used for natural drying.

Dried fruit and vegetables have certain advantages over those preserved by other methods. They are lighter in weight than their corresponding fresh produce and, at the same time, they do not require refrigerated storage. However, if they are kept at high temperatures and have a high moisture content they will turn brown after relatively short periods of storage. Different lots at various stages of maturity (ripeness) must NOT be mixed together; this would result in a poor dried product. Some varieties of fruit and vegetables are better for natural drying than other; they must be able to withstand natural drying without their texture becoming tough so that they are not difficult to reconstitute. Some varieties are unsuitable because they have irregular shape and there is a lot of wastage in trimming and cutting such varieties.

Damaged parts which have been attacked by insects, rodents, diseases, etc. and parts which have been discoloured or have a bad appearance or colour, must be removed. Before trimming and cutting, most fruit and vegetables must be washed in clean water. Onions are washed after they have been peeled.

Trimming includes the selection of the parts which are to be dried, cutting off and disposing of all unwanted material. After trimming, the greater part of the fruit and vegetables cut into even slices of about 3 to 7 mm thick or in halves/quarters, etc.

It is very important to have all slices/parts in one drying lot of the same thickness/size; the actual thickness will depend on the kind of material. Uneven slices or different sizes dry at different rates and this result in a poor quality end product. Onions and root crops are sliced with a hand slicer or vegetable cutter; bananas, tomatoes and other vegetables or fruit are sliced with stainless-steel knives.

As a general rule plums, grapes, figs, dates are dried as whole fruits without cutting/slicing.

Some fruit and vegetables, in particular bananas, apples and potatoes, go brown very quickly when left in the air after peeling or slicing; this discolouration is due to an active enzyme called phenoloxidase. To prevent the slices from going brown they must be kept under water until drying can be started. Salt or sulphites in solution give better protection. However, whichever method is used, further processing should follow as soon as possible after cutting or slicing.

Blanching - exposing fruit and vegetable to hot or boiling water - as a pre-treatment before drying has the following advantages:

- It helps clean the material and reduce the amount of micro-organisms present on the surface;
- It preserves the natural colour in the dried products; for example, the carotenoid (orange and yellow) pigments dissolve in small intracellular oil drops during blanching and in this way they are protected from oxidative breakdown during drying;
- It shortens the soaking and/or cooking time during reconstitution.

During hot water blanching, some soluble constituents are leached out: water-soluble flavours, vitamins (vitamin C) and sugars. With potatoes this may be an advantage as leaching out of sugars makes the potatoes less prone to turning brown.

Blanching is a delicate processing step; time, temperature and the other conditions must be carefully monitored.

A suitable water-blanching method in traditional processing is as follows:

- The sliced material is placed on a square piece of clean cloth; the corners of the cloth are tied together;
- A stick is put through the tied corners of the cloth;
- The cloth is dipped into a pan containing boiling water and the stick rests across the top of the pan thus providing support for the cloth bag.

The average blanching time is 6 minutes. The start of blanching has to be timed from the moment the water starts to boil again after the cloth bag has been dipped into the pan. While the material is being blanched the cloth bag should be raised and lowered in the water so that the material is heated evenly.

When the blanching time is completed the cloth bag and its content should be dipped into cold water to prevent over-blanching. If products are over-blanched (boiled for too long) they will stick together on the drying trays and they are likely to have a poor flavour.

Green beans, carrots, okra, turnip and cabbage should always be blanched. The producer can choose whether or not potatoes need blanching. Blanching is not needed for onions, leeks, tomatoes and sweet peppers. Tomatoes are dipped into hot water for one minute when they need to be peeled but this is not blanching.

Use of Preservatives

Preservatives are used to improve the colour and keeping qualities of the final product for some fruits and vegetables. Preservatives include items such as sulphur dioxide, ascorbic acid, citric acid, salt and sugar and can either be simple or compound solutions.

Treatment with preservatives takes place after blanching or, when blanching is not needed, after slicing. In traditional, simple processing the method recommended is:

- Put enough preservative solution to cover the cloth bag into a container/pan;
- Dip the bag containing the product into the preservative solution for the amount of time specified;
- Remove the bag and put it on a clean tray while the liquid drains out. The liquid which drains out must not go back into the preservative solution because it would weaken the solution.

Care must be taken after each dip to refill the container to the original level with fresh preservative solution of correct strength. After five lots of material have been dipped, the remaining solution is thrown away; i.e. a fresh lot of preservative solution is needed for every 5 lots of material. The composition and strength of the preservative solution vary for different fruit and vegetables.The strength of sulphur dioxide is expressed as "parts per million" (ppm). 1.5 grams of sodium metabisulphite in one litre of water gives 1000 ppm of sulphur dioxide. Details for solutions of different strengths are given in the following table.

Table. Dilutions of sodium metabisulphite with water to obtain "PM" of sulphur dioxide (SO2)

PPM SO_2	SODIUM METABISULPHITE	
	Grams per litre of	**Grams per 20 litre tin**
water		**of water**
1000	1.5	30.0
2000	3.0	60.0
3000	4.5	90.0
4000	6.0	120.0
5000	7.5	150.0
6000	9.0	180.0
7000	10.5	210.0

One level teaspoon of sodium metabisulphite = c. 5 g.

The preservative solutions in the fruit and vegetable pre-treatment can only be used in enamelled, plastic or stainless-steel containers; never use ordinary metal because solutions will corrode this type of container.

As a general rule, preservatives are not used for treating onions, garlic, leeks, chilies and herbs.

Osmotic Dehydration

In osmotic dehydration the prepared fresh material is soaked in a heavy (thick liquid sugar solution) and/or a strong salt solution and then the material is sun or solar dried. During osmotic treatment the material loses some of its moisture. The syrup or salt solution has a protective effect on colour, flavour and texture.This protective effect remains throughout the drying process and makes it possible to produce dried products of high quality. This process makes little use of sulphur dioxide.

Sun Drying

The main problems for sun drying are dust, rain and cloudy weather. Therefore, drying areas should be dust-free and whenever there is a threat of a dust storm or rain, the drying trays should be stacked together and placed under cover.

In order to produce dust-free and hygienically clean products, fruit and vegetable material should be dried well above ground level so that they are not contaminated by dust, insects, livestock or people. All materials should be dried on trays designed for the purpose; the most common drying trays have wooden frames with a fitted base of nylon mosquito netting. Mesh made of woven grass can also be used. Metal netting must NOT be used because it discolours the product.The trays should be placed on a framework at table height from the ground. This allows the air to circulate freely around the drying material and it also keeps the food product well away from dirt. Ideally the area should be exposed to wind and this speed up drying, but this can only be done if the wind is free of dust.With 80 cm x 50 cm trays, the approximate load for a tray is 3 kg; the material should be spread in even layers. During the first part of the drying period, the material should be stirred and turned over at least once an hour.

This will help the material dry faster and more evenly, prevent it sticking together and improve the quality of the finished product. Products for sun drying should be prepared early in the day; this will ensure that the material enjoys the full effect of the sun during the early stages of drying.

At night the trays should be stacked in a ventilated room or covered with canvas. Plastic sheets should NEVER be used for covering individual trays during sun drying.Dry or nearly dry products can be blown out of the tray by the wind. However, this can be protected by covering the loaded tray with an empty one; this also gives protection against insects and birds.

Shade Drying

Shade drying is carried out for products which can lose their colour and/or turn brown if put in direct sunlight. Products which have naturally vivid colours like herbs, green and red sweet peppers, chilies, green beans and okra give a more attractive end-product when they are dried in the shade.

The principles for the shade drying are the same as for sun drying. The material to be dried requires full air circulation. Therefore, shade drying is carried out under a roof or thatch which has open sides; it cannot be done either inside conventional buildings with side walls or in compounds sheltered from wind. Under dry conditions when there is a good circulation of air, shade drying takes little more time than is normally required for drying in full sunlight.

Identification of Suitable Designs of Solar Dryers for Different Applications

In the selection of appropriate solar dryers for commercial scale operation, it is imperative that economics be kept in view at all time. A total Energy System concept should be employed and due consideration be given to parasitic energy consumption.

The following features have been identified:

- Large scale dryers are more promising than small scale ones. However, small scale dryers should not be neglected.
- The dryer should be designed to maximise the utilisation factor of the capital investment, i.e. multi-products (fruit, vegetables and other raw material) and multi-use (e.g. drying and heating water for domestic use).
- In general, an auxiliary heat source should be provided to assure

reliability, to handle peak loads and also to provide continuous drying during periods of no sunshine.

- Forced convection indirect dryers are preferred because they offer better control, more uniform drying and because of their high heat collection efficiency result in smaller collector area. However, parasitic power should be kept to a minimum.

Two dryer systems have been identified:

- A cabinet type dryer with natural convection for internal air circulation for the processing of dried fruit such as mango, banana, pineapple, apricot, pear, apple, etc. and also for potato chips and other vegetables;
- A greenhouse type dryer with forced air circulation.

Some of the barriers to the commercial development of solar dryers have been attributed to:

- Initial cost - poor farmers cannot afford them
- Durability - constant breakdown due to using low cost building materials
- Misuse - through lack of training and technical skills
- Dependability and reliability - during the wet season when drying is critical there is not enough solar energy available
- The wider use of Solar Drying Systems has been limited by other factors which are not necessarily of a technical or technological nature. Among the most important are the lack of national policies directed to promoting the drying of produce at the production site, in order to reduce losses, improve quality and increase farmers' earnings.

Construction of Solar Dryers

In the case of simple natural convection dryers it may be more appropriate to build and operate a number of small units. Multiplicity allows diversity, since more than one crop can be dried at a time. A further advantage is that if one dryer is out of operation due to damage, drying can still continue at reduced capacity using the other dryers.

On the other hand, more sophisticated dryers, such as forced convection solar dryers, benefit from economies of scale due to the investment tied up in the fan and the source of heat.Generally speaking, one large dryer will be more cost-effective than two smaller units. However, it should be taken into consideration that an oversized unit will be operating at less than full capacity, reducing any cost advantage. The drying area required will depend on local conditions, commodity and number of trays on each rack or trolley.

Construction Methods and Materials

Construction methods and available materials may vary considerably from location to location. It is not within the scope of this document to discuss individual, local circumstances. Some general guidelines regarding factors which must be considered can, however, be given:

— *Dimensions of standard materials:* Where possible, design should take account of the sizes of material locally available. For example, it would be poor design to specify the width of a corrugated iron collector as 1.1 m if the standard width of a corrugated iron sheet is 1 m.

Before finalising a design the commercial availability of materials must be ascertained.

— *Use of rural materials:* The cost of building of solar dryer can be minimised if the producer is able to use wood cut straight from the forest rather than prepared timber.

Careful design in the development stage of a dryer can often facilitate the use of cheaper materials. Difficulties caused by these materials are in joining pieces of the structure, in sealing the structure against air leaks, and in attaching the plastic sheet to the (wooden) frame. There is obvious scope for designs which use prepared timber for strategic points and unprepared at others.

Where the use of wood is necessary, remember to take environmental factors into consideration. For example, determine the effect of flash flooding or termites might be and take the appropriate preventive action.

— *Use of plastic sheets:* For many solar dryers, the clear plastic sheet used is the major capital cost to the farmer; therefore, the type of plastic chosen is important.

A choice must be made between a relatively cheap plastic such as ordinary polyethylene which will last, at best, for one season due to photo-degradation and wear and tear; and a more expensive, better quality plastic less prone to photo-degradation; or even glass or a rigid plastic. Attaching plastic sheet to the framework structure, so as to minimise the likelihood of the plastic being torn is, perhaps, the most difficult part of building a dryer. Listed below are some general points which should be followed to prolong the useful lifetime of plastic sheet on a solar dryer:

- When attaching plastic sheet to the framework, care should be taken to stretch the plastic at the points of attachment, but the plastic should not be so loose that it will flap about in the wind;
- Rather than merely stapling or nailing the plastic directly to the framework, it is preferable to sandwich the plastic between the framework and a batten. This may not be practical when unprepared wood or other materials are being used.
- No sharp edges should come in contact with the plastic sheet since these will initiate tears;
- Fold over the plastic at the point of attachment to the frame, so that there are two or more layers of plastic. This will help prevent tears;
- When fixing the sheet over the framework, sags and hollows in which water can collect should be avoided wherever possible;

- The dryer should be handled as carefully and as seldom as possible during operation and when not in use.

Technical Criteria

The following design factors must be established:

- The throughput of the dryer over the productive season; the size of batch to be dried;
- The drying period(s) under stated conditions;
- The initial and desired final moisture content of the commodity (if known);
- The drying characteristics of the commodity, such as maximum drying temperature, effect of sunlight upon the product quality, etc.;
- Climatic conditions during the drying season, i.e. sunlight intensity and duration; air temperature and humidity; wind speed (such data may be available from local meteorological stations);
- Availability and reliability of electrical power;
- The availability, quality, durability and price of potential construction materials such as:
  - — Glazing materials: glass, plastic sheet or film;
  - — Wood (prepared or unprepared);
  - — Nails, screw, bolts, etc.;
  - — Metal sheet, flat or corrugated angle iron;
  - — Bricks (burnt or mud), concrete blocks, stones, cement, sand, etc.
  - — Roofing thatch;
  - — Metal mesh, wire netting, etc.
  - — Mosquito netting, muslin, etc.
  - — Bamboo or fibre weave;
  - — Black paint, other blackening materials;
  - — Insulation material; sawdust, etc.;
- The type of labour available to build and operate the dryer;
- The availability of clean water at the site for preparation of the commodity prior to drying.

In any one situation there may well be other technical factors that need to be considered.

Socio-economic Criteria

From the initial considerations, estimates of the capital costs of the dryer, the price of the commodity to be dried, and the likely selling price of the dried product will have been made. Other question that need to be considered are the following:

- Who will own the dryer?

- Is the dryer to be constructed by the end-user (with or without advice from extension agencies), local contractors, or other organisations?
- Who will operate and maintain it?
- How can the drying operation be incorporated into current practices?
- Are sources of finance from local authorities or extension agencies available, etc.?

Obviously there are many other socio-economic factors, particularly those of a local nature, which must be taken into account. It cannot be stressed too highly that if such factors are not taken into account and evaluated, then is every chance that an inappropriate dryer design may result. Equal emphasis must be placed on both technical and socio-economic factors.

a. Situations where solar dryer may be useful:
   - Where the cost of conventional energy is prohibitive and/or the supply is erratic, to supplement existing artificial drying systems and reduce fuel costs;
   - Where land is in short supply or expensive;
   - Where the quality of existing sun dried products can be improved upon;
   - Where the labour is in short supply;
   - Where is plenty of sunshine, but high humidity.
b. Situations where solar dryers may not be useful:
   - Where conventional energy sources are abundant and cheap;
   - Where large amounts of combustible by-products or waste materials are freely available;
   - Where there is insufficient sunshine;
   - Where is plenty of sunshine and arid conditions (sun drying may suffice);
   - Where the quality of sun dried products already made cannot be improved upon;
   - Where local operators are insufficiently trained;
   - Where the ramifications of introducing a solar dryer have not been completely thought out.

Sun/solar Drying Tray

The drying tray described requires seasoned timber 22.5 mm thick x 50 mm wide. A sun drying tray requires 6 metres of seasoned timber 22.5 mm thick x 50 mm wide.

Cut the timber into lengths of 900 mm long for the sides of the tray and 600 mm long for the ends - 4 pieces of each length will be needed. The ends of each piece are cut as shown in the drawing - this is to make flush fitting joints. Join the corners using small brass screws 20 mm long. To make extra strong joints use good quality wood glue as well as the screws.The nylon mosquito netting or grass woven mesh can be fitted between the frames as shown in the bottom drawing. Cut the mesh a little larger than the size of the frame. Using drawing pins, pin the mesh to the OUTSIDE edges of one of the frames - the mesh should be pulled tight as the pins are put in around the edges.

Lay the other frame on top and drill holes about 3 mm in diametre at the points marked X in the top drawing. Use nails that are a tight fit in the holes and tap gently into place leaving a portion standing above the frame.

Cut off the standing part to leave a piece about 12 mm long which is then bent over and tapped firmly down onto the frame. When the frame has been put together tightly, the drawing pins can be removed.

Preservation by Concentration

Foods are concentrated for many of the same reasons that they are dehydrated; concentration can be a form of preservation but this is true only for some foods. Concentration reduces weight and volume and results in immediate economic advantages.

Nearly all liquid foods which are dehydrated are concentrated before they are dried. This is because in the early stages of water removal, moisture can be more economically removed in highly efficient evaporators than in dehydration equipment. Further, increased viscosity from concentration often is needed to prevent liquids from running off drying surfaces or to facilitate foaming or puffing. Foods are also concentrated because the concentrated forms have become desirable components of diet in their own right. Thus, fruit juices plus sugar with concentration becomes jelly. The more common concentrated fruit and vegetable products include items as fruit and vegetable juices and nectars, jams and jellies, tomato paste, many types of fruit purées used by bakers, candy makers and other food manufacturers.

Aspects of Preservation by Concentration

The level of water in virtually all concentrated foods is in itself more than enough to permit microbial growth. Yet while many concentrated foods such as non-acid fruit and vegetable purées may quickly undergo microbial spoilage unless additionally processed, such items as sugar syrups, jellies and jams are relatively "immune" to spoilage; the difference of course is in what is dissolved in the remaining water and what osmotic concentration is reached.

1. Cell walls
2. Trays on a car
3. Metal plat with burning sulphur
4. Hole for sulphur dioxide fumes
5. Exhaust hole

Courtesy of U.T.A. Industrie

1. Control and switch-board

2. Burner
3. Platform for burner
4. Frontal plate
5. Air flow regulating plate
6. Burner's cylinder
7. Air circulating fan
8. Air direction conveying plates
9. Cars rail
10. Tunnel "feeding" door: inlet of cars with fresh product
11. Tunnel "evacuating" door: outlet of cars with dry product
12. Cars pushing device
13. Cars with trays
14. Drying trays

Sugar and salt in concentrated solutions have high osmotic pressure. When these are sufficient to draw water from microbial cells or prevent normal diffusion of water into these cells, a preservative condition exists.The critical concentration of sugar in water to prevent microbial growth will vary depending upon the type of micro-organisms and the presence of other food constituents, but usually 70% sucrose in solution will stop growth of all micro-organisms in foods. Less than this concentration may be effective but for short periods of time unless the foods contain acid or they are refrigerated.

Salt becomes a preservative when its concentration is increased and levels of about 18% to 25% in solution generally will prevent all growth of micro-organisms in foods. Except in the case of certain briny condiments, however, this level is rarely tolerated in foods.Removal of water by concentration also increases the level of food acids in solution (particularly significant in concentrated fruit juices).

Reduced Weight and Volume by Concentration

While the preservation effects of food concentration are important, the main reason of most food concentration is to reduce food weight and bulk. Tomato pulp which is ground tomato minus the skins and seeds, has a solid content of only 6 % and so a 3.785 litre can would contain only 231 g of tomato solids.

Table: Specific gravity and solids. Tomato pulp and commercial tomato concentrates

Tomato Solids, %	Specific gravity at 68°F (20°C)	Dry Tomato Solids	
		per Gal. at 68°F, lb	**per Litre at 20°C, g**
6.0 Tomato pulp	1.025	0.51	61
10.8	1.045	0.94	113
12.0 Tomato purée	1.050	1.05	126
14.2	1.060	1.25	151

16.5	1.070	1.47	177
25.0	1.107	2.31	277
26.0	1.112	2.41	289
28.0 Tomato paste	1.120	2.61	314
30.0	1.129	2.82	339
32.0	1.138	3.03	364

Concentrated to 32% solids, the same can would contain 1.38 kg of tomato solids or six times the value of product.

For a manufacturer needing tomato solids such a producer of soups, canned spaghetti or frozen pizza the saving from concentration are enormous in cans, transportation costs, warehousing costs and handling costs throughout his operation.

Methods of Concentration

a. Solar concentration. As in food dehydration, one of the simplest methods of evaporating water is with solar energy. A typical example of this method is production at farm level in developing countries of fruit pastes/leathers (such as apricot or plum pastes).

b. Open Kettles. Some foods can be satisfactorily concentrated in open kettles that are heated by steam. This is the case for jellies and jams, tomato juices and purées and for certain types of soups. High temperatures and long concentration times should be avoided in order to reduce or eliminate damage.

   It is also necessary to avoid thickening and burn-on of product to the kettle wall as these gradually lower the efficiency of heat transfer and slow the concentration process.

   However, when the process is under control, this type of evaporation is still highly recommended for small scale operations in developing countries. It is a quite widely used system, mainly for jellies, jams and marmalades.

c. Flash evaporators.

d. Thin-film Evaporators.

e. Vacuum evaporators. - It is common to construct several vacuumised vessels in series so that the product moves from one vacuum chamber to the next and thereby becomes progressively more concentrated in stages.

   With such an arrangement the successive stages are maintained at progressively higher degrees of vacuum, and the hot water vapour produced by the first stage is used to heat the second stage, the vapour from the second stage heats the third stage and so on. In this way maximum use of heat energy is made.

f. Freeze Concentration. This process has been known for many years and has been applied commercially to orange juice. However, high processing costs due largely to losses of juice occlude [unclear] to the ice crystals, have limited the number of installations to date.

g. Ultrafiltration and reverse Osmosis.

Changes from Concentration

Obviously concentration that exposes food to 100° C or higher temperatures for prolonged periods can cause major changes in organoleptic and nutritional properties. Cooked foods and darkening of colour are two of the more common heat induced results which must be kept under control during a well designed process with an efficient evaporator which is still "safe".Microbial destruction is another type of change that may occur during concentration and will be largely dependent upon temperature. Concentration at a temperature of 100 °C or slightly above will kill many micro-organisms but cannot be depended on to destroy bacterial spores. When the food contains acid, such as fruit juices, the kill will be greater but again sterility is unlikely.

On the other hand, when concentration is done under vacuum many bacterial types not only survive the low temperatures but multiply in the concentrating equipment. It is therefore necessary to stop frequently and sanitise low temperature evaporators and where sterile concentrated foods are required, to resort to an additional preservation treatment.

Many chemicals will kill micro-organisms or stop their growth but most of these are not permitted in foods; chemicals that are permitted. Chemical food preservatives are those substances which are added in very low quantities (up to 0.2%) and which do not alter the organoleptic and physico-chemical properties of the foods at or only very little.

Preservation of food products containing chemical food preservatives is usually based on the combined or synergistic activity of several additives, intrinsic product parametres (e.g. composition, acidity, water activity) and extrinsic factors (e.g. processing temperature, storage atmosphere and temperature).

This approach minimises undesirable changes in product properties and reduces concentration of additives and extent of processing treatments.

The concept of combinations of preservatives and treatments to preserve foods is frequently called the hurdle or barrier concept. Combinations of additives and preservatives systems provide unlimited preservation alternatives for applications in food products to meet consumer demands for healthy and safe foods.

Chemical food preservatives are applied to foods as direct additives during processing, or develop by themselves during processes such as fermentation. Certain preservatives have been used either accidentally or intentionally for centuries, and include sodium chloride (common salt), sugar, acids, alcohols and components of smoke. In addition to preservation, these compounds contribute to the quality and identity of the products, and are applied through processing procedures such as salting, curing, fermentation and smoking.

Traditional chemical food preservatives and their use in fruit and vegetable processing technologies could be summarised as follows: common salt: brined vegetables; sugars (sucrose, glucose, fructose and syrups): foods preserved by high sugar concentrations: jellies, preserves, syrups, juice concentrates; interaction of sugar with other ingredients or processes such as drying and heating; indirect food preservation by sugar in products where fermentation is important (naturally acidified pickles and sauerkraut).

Acidulants and other preservatives formed in or added to fruit and vegetable products are as follows:

Lactic acid. This acid is the main product of many food fermentations; it is formed by microbial degradation of sugars in products such as sauerkraut and pickles. The acid produced in such fermentations decreases the pH to levels unfavourable for growth of spoilage organisms such as putrefactive anaerobes and butyric-acid-producing bacteria. Yeasts and moulds that can grow at such pH levels can be controlled by the inclusion of other preservatives such as sorbate and benzoate.Acetic acid. Acetic acid is a general preservative inhibiting many species of bacteria, yeasts and to a lesser extent moulds. It is also a product of the lactic-acid fermentation, and its preservative action even at identical pH levels is greater than that of lactic acid. The main applications of vinegar (acetic acid) includes products such as pickles, sauces and ketchup.

Other acidulants

- Malic and tartaric (tartric) acids is used in some countries mainly to acidify and preserve fruit sugar preserves, jams, jellies, etc.
- Citric acid is the main acid found naturally in citrus fruits; it is widely used (in carbonated beverages) and as an acidifying agent of foods because of its unique flavour properties. It has an unlimited acceptable daily intake and is highly soluble in water. It is a less effective antimicrobial agent than other acids.
- Ascorbic acid or vitamin C, its isomer isoascorbic or erythorbic acid and their salts are highly soluble in water and safe to use in foods.

Commonly Used Lipophilic Acid Food Preservatives

Benzoic acid in the form of its sodium salt, constitutes one of the most common chemical food preservative. Sodium benzoate is a common preservative in acid or acidified foods such as fruit juices, syrups, jams and jellies, sauerkraut, pickles, preserves, fruit cocktails, etc. Yeasts are inhibited by benzoate to a greater extent than are moulds and bacteria.

Sorbic acid is generally considered non toxic and is metabolised; among other common food preservatives the WHO has set the highest acceptable daily intake (25 mg/kg body weight) for sorbic acid.Sorbic acid and its salts are practically tasteless and odourless in foods, when used at reasonable levels (< 0.3 %) and their antimicrobial activity is generally adequate.

Sorbates are used for mould and yeast inhibition in a variety of foods including fruits and vegetables, fruit juices, pickles, sauerkraut, syrups, jellies, jams, preserves, high moisture dehydrated fruits, etc.

Potassium sorbate, a white, fluffy powder, is very soluble in water (over 50%) and when added to acid foods it is hydrolysed to the acid form. Sodium and calcium sorbates also have preservative activities but their application is limited compared to that for the potassium salt, which is employed because of its stability, general ease of preparation and water solubility.

Gaseous Chemical Food Preservatives

Sulphur dioxide and sulphites. Sulphur dioxide (SO_2) has been used for many centuries as a fumigant and especially as a wine preservative. It is a colourless, suffocating, pungent-smelling, non-flammable gas and is very soluble in cold water (85 g in 100 ml at 25°C).

Sulphur dioxide and its various sulphites dissolve in water, and at low pH levels yield sulphurous acid, bisulphite and sulphite ions. The various sulphite salts contain 50-68% active sulphur dioxide. A pH dependent equilibrium is formed in water and the proportion of SO_2 ions increases with decreasing pH values. At pH values less than 4.0 the antimicrobial activity reaches its maximum.

Sulphur dioxide is used as a gas or in the form of its sulphite, bisulphite and metabisulphite salts which are powders. The gaseous form is produced either by burning Sulphur or by its release from the compressed liquefied form.

Metabisulphite are more stable to oxidation than bisulphites, which in turn show greater stability than sulphites. The antimicrobial action of sulphur dioxide against yeasts, moulds and bacteria is selective, with some species being more resistant than others.

Sulphur dioxide and sulphites are used in the preservation of a variety of food products. In addition to wines these include dehydrated/dried fruits and vegetables, fruit juices, acid pickles, syrups, semi-processed fruit products, etc. In addition to its antimicrobial effects, sulphur dioxide is added to foods for its antioxidant and reducing properties, and to prevent enzymatic and non-enzymatic browning reactions.

Carbon dioxide (CO_2) is a colourless, odourless, non-combustible gas, acidic in odour and flavour. In commercial practice it is sold as a liquid under pressure (58 kg per cm^3) or solidified as dry ice.

Carbon dioxide is used as a solid (dry ice) in many countries as a means of low-temperature storage and transportation of food products. Beside keeping the temperature low, as it sublimes, the gaseous CO_2 inhibits growth of psychrotrophic micro-organisms and prevents spoilage of the food (fruits and vegetables, etc.).Carbon dioxide is used as a direct additive in the storage of fruits and vegetables. In the controlled/modified environment storage of fruit and vegetables, the correct combination of O_2 and CO_2 delays respiration and ripening as well as retarding mould and yeast growth.

The final result is an extended storage of the products for transportation and for consumption during the off-season. The amount of CO_2 (5-10%) is determined by factors such as nature of product, variety, climate and extent of storage.Chlorine. The various forms of chlorine constitute the most widely used chemical sanitiser in the food industry. These chlorine forms include chlorine (Cl_2), sodium hypochlorite (NaOCl), calcium hypochlorite ($Ca(OCl)_2$) and chlorine dioxide gas (ClO_2).

These compounds are used as water adjuncts in processes such as product washing, transport, and cooling of heat-sterilised cans; in sanitising solutions for equipment surfaces, etc.Important applications of chlorine and its compounds include disinfection of drinking water and sanitation of food processing equipment.

General Rules for Chemical Preservation

Chemical food preservatives have to be used only at a dosage level which is needed for a normal preservation and not more. "Reconditioning" of chemical preserved food, e.g. a new addition of preservative in order to stop a microbiological deterioration already occurred is not recommended.

The use of chemical preservatives must be strictly limited to those substances which are recognised as being without harmful effects on human beings' health and are accepted by national and international standards and legislation.

Factors which Determine/Influence the Action of Chemical Food Preservatives

Factors related to the chemical preservatives:

a. Chemical composition;
b. Concentration.

Factors related to micro-organisms:

a. Micro-organism species; as a general rule it is possible to take the following facts as a basis:
   - Sulphur dioxide and its derivatives can be considered as an "universal" preservative; they have an antiseptic action on bacteria as well as on yeasts and moulds;
   - Benzoic acid and its derivatives have a preservative action which is stronger against bacteria than on yeasts and moulds;
   - Sorbic acid acts on moulds and certain yeast species; in higher dosage levels it acts also on bacteria, except lactic and acetic ones;

- Formic acid is more active against yeasts and moulds and less on bacteria.

b. The initial number of micro-organisms in the treated product determines the efficiency of the chemical preservative.

The efficiency is less if the product has been contaminated because of preliminary careless hygienic treatment or an incipient alteration. Therefore, with a low initial number of micro-organisms in the product, the preservative dosage level could be reduced.

Specific factors related to the product to be preserved:

a. Product chemical composition;
b. Influence of the pH value of the product: the efficiency of the majority of chemical preservatives is higher at lower pH values, i.e. when the medium is more acidic.
c. Physical presentation and size which the product is sliced to: the chemical preservative's dispersion in food has an impact on its absorption and diffusion through cell membranes on micro-organisms and this determines the preservation effect.

Therefore, the smaller the slicing of the product, the higher the preservative action. Preservative dispersion is slowed down by viscous foods (concentrated fruit juices, etc.)

Miscellaneous factors

a. *Temperature:* chemical preservative dosage level will be established as a function of product temperature and characteristics of the micro-flora;
b. *Time:* at preservative dosage levels in employed in industrial practice, the time period needed in order to obtain a "chemical sterilisation" is a few weeks for benzoic acid and shorter for sulphurous acid.

Usual accepted chemical food preservatives are detailed in Table given below.

Table: Chemical Food Preservatives

Agent	Acceptable Daily intake (mg/Kg body weight)	Commonly used levels (%)
Lactic acid	No limit	No limit
Citric acid	No limit	No limit
Acetic acid	No limit	No limit
Sodium Diacetate	15	0.3-0.5
Sodium benzoate	5	0.03-0.2
Sodium propionate	10	0.1-0.3
Potassium sorbate	25	0.05-0.2
Methyl paraben	10	0.05-0.1

Sodium nitrite	0.2	0.01-0.02
Sulphur dioxide	0.7	0.005-0.2

For the purpose of this document, some food products in common usage are summarised as follows:

- Citric acid: fruit juices; jams; other sugar preserves;
- Acetic acid: vegetable pickles; other vegetable products;
- Sodium benzoate: vegetable pickles; preserves; jams; jellies; semi-processed products;
- Sodium propionate: fruits; vegetables;
- Potassium sorbate: fruits; vegetables; pickled products; jams, jellies;
- Methyl paraben: fruit products; pickles; preserves;
- Sulphur dioxide: fruit juices; dried/dehydrated fruits and vegetables; semi-processed products.

PRESERVATION OF VEGETABLES BY ACIDIFICATION

Food acidification is a means of preventing their deterioration in so far as a non-favourable medium for micro-organisms development is created. This acidification can be obtained by two ways: natural acidification and artificial acidification.

Natural Acidification

This is achieved by a predominant lactic fermentation which assures the preservation based on acidoceno-anabiosys principle; preservation by lactic fermentation is called also biochemical preservation. Throughout recorded history food has been preserved by fermentation. In spite of the introduction of modern preservation methods, lactic acid fermented vegetables still enjoy a great popularity, mainly because of their nutritional and gastronomic qualities.

The various preservation methods discussed thus far, based on the application of heat, removal of water, cold and other principles, all have the common objective of decreasing the number of living organisms in foods or at least holding them in check against further multiplication.

Fermentation processes for preservation purposes, in contrast, encourage the multiplication of micro-organisms and their metabolic activities in foods. But the organisms that are encouraged are from a select group and their metabolic activities and end products are highly desirable. The extent of this desirability is emphasised by a partial list of fermented fruits and vegetable products from various parts of the world.

There are some characteristic features in the production of fermented vegetables which will be pointed out below using cucumbers as an example. In the production of lactic acid fermented cucumbers, the raw material is put into a brine without previous heating. Through the effect of salt and oxygen deficiency the cucumber tissues gradually die. At the same time, the semipermeability of the cell membranes is lost, whereby soluble cell components diffuse into the brine and serve as food substrate for the micro-organisms.

Under such specific conditions of the brine the lactic acid bacteria succeed in overcoming the accompanying micro-organisms and lactic acid as the main metabolic products is formed. Under favourable conditions (for example moderate salt in the brine, use of starter cultures) it takes at least 3 days until the critical pH value of 4.1 or less - desired for microbiological reasons - is reached.Beside the typical taste, for the consumer a crisp texture is the most important quality criterion for fermented vegetables. Because there is no heating step before the fermentation, the indigenous plant enzymes in the fermenting materials are still present during the very first phase. After the destruction of the cell membranes they easily get to their active sites and under favourable conditions they can easily cause softening.

The environmental conditions act in a different manner on single enzymes or enzymes systems: some enzymes are strongly inhibited by salt, others are activated, and in the acid pH-region many enzymes are irreversibly inactivated. Beside indigenous enzymes also enzymes produced by micro-organisms can be responsible for the undesired soft products.

In technically advanced societies the major importance of fermented foods has come to be variety they add to the diet. However, in many less developed areas of the world, fermentation and natural drying are the major food preservation methods and as such are vital to survival of a large proportion of the world's current population.

Artificial acidification is carried out by adding acetic acid which is the only organic acid harmless for human health and stable in specific working conditions; in this case biological principles of the preservation are acidoanabiosys and, to a lesser extent, acidoabiosys.

Combined acidification is a preservation technology which involves as a preliminary processing step a weak lactic fermentation followed by acidification (vinegar addition). The two main classes of vegetables preserved by acidification are sauerkraut and pickles; the definitions of these products adapted from US Code of Federal Register are as follows.

Bulk sauerkraut. Bulk or barrelled sauerkraut is the product of characteristic acid flavour, obtained by the full fermentation, chiefly lactic, of properly prepared and shredded cabbage in the presence of 2-3% salt. On completion of fermentation, it contains not more than 1.5% of acid, expressed as lactic acid.

Canned sauerkraut. Canned (or packaged) sauerkraut, is prepared from clean, sound, well-matured heads of the cabbage plant (Brassica oleracea var. capitata L.) which have been properly trimmed and cut; to which salt is added and which is cured by natural fermentation. The product may or may not be packed with pickled peppers, pimientos, or tomatoes or contain other flavouring ingredients to give the product specific flavour characteristics. The product:

a. May be canned by processing sufficiently by heat to assure preservation in hermetically sealed containers; or
b. May be packaged in sealed containers and preserved with or without the addition of benzoate of soda or any other ingredient permissible under the provisions of Food and Drug Administration (FDA).

Pickles. "Pickles" means the product prepared entirely or predominantly from cucumbers (Cucumis sativus L.). Clean, sound ingredients are used which may or may not have been previously subjected to fermentation and curing in a salt brine (solution of sodium chloride, NaCl).

The prepared pickles are packed in a vinegar solution to which may be added salt and other vegetables, nutritive sweeteners, seasonings, flavourings, spices, and other ingredients permissible under FDA regulations. The product is packed in suitable containers and heat treated, or otherwise processed to assure preservation. Sauerkraut and pickle products can be preserved under the effect of natural or added acidity, followed by pasteurization when this acidification is not sufficient.

Sauerkraut is a very good source of vitamin C; the importance of this product should be emphasised in developing countries as a simple technology which can be applied mainly for consumption of the finished products in remote, isolated areas during the cold season. It is also a excellent technology to be learned to schools which have their own source of cabbage and cucumbers through school agricultural farms.

Sauerkraut and pickles are manufactured on an industrial scale in significant quantities world-wide. However, the basic technology is simple and could be applied at home, farm and community level after some explanation and training. The natural acidification preservation could be considered similar to sun/solar drying in terms of training and development.

Table. Some industrial fermentation processes in food industries

I. *Lactic acid bacteria*	
— cucumbers	dill pickles, sour pickles
— cabbage	sauerkraut
— turnips	sauerruben
— lettuce	lettuce kraut
— mixed vegetables, turnips, radish, cabbage	
— mixed Chinese vegetables, cabbage	Kimchi
— vegetables and milk	Tarhana
— vegetables and rice	Sajur asin
II. *Lactic acid bacteria with other micro-organisms*	
— with yeasts	Nukamiso pickles
— with moulds	tempeh, soy sauce

III. Acetic acid bacteria - wine, cider or any alcoholic and sugary or starchy products may be converted to vinegar

IV. Yeasts

– fruit	wine, vermouth

The principle of this technology is to add sugar in a quantity that is necessary to augment the osmotic pressure of the product's liquid phase at a level which will prevent microorganism development.

From a practical point of view, however, it is usual to partially remove water (by boiling) from the product to be preserved, with the objective of obtaining a higher sugar concentration. In concentrations of 60% in the finished products, the sugar generally assures food preservation.

It is important to know the ratio between the total sugar quantity in the finished product and the total sugar concentration in the liquid phase because this determines, in practice, the sugar preserving action.

The percent composition of a product preserved with sugar, for example marmalade, can be expressed as follows: $[i + S + s + n + w] = 100$;

i = insoluble substance;

s = sugar from fruits;

S = added sucrose;

n = soluble "non sugar"

w = water.

In this case, total sugar concentration, in the liquid phase, of the finished product is:

$$X = 100 (S + s)/100 - (n + i) [\%]$$

Therefore, in the case of a standard marmalade with 55 % sugar added (calculated on the finished product basis), the real concentration in the liquid phase is for example:

$$X = 100 (55 + 8)/100 - (5 + 3) = 68.5\%$$

In the food preservation with sugar, the water activity cannot be reduced below 0.845; this value is sufficient for bacteria and neosmophile yeast inhibition but does not prevent mould attack. For this reason, various means are used to avoid mould development:

- Finished product pasteurization (jams, jellies, etc.);
- Use of chemical preservatives in order to obtain the antiseptisation of the product surface.

It is very important from a practical point of view to avoid any product contamination after boiling and to assure an hygienic operation of the whole technological process (this will contribute to the prevention of product moulding or fermentation).

Storage of the finished products in good conditions can only be achieved by ensuring the above level of water activity.

HEAT PRESERVATION/HEAT PROCESSING

Various Degrees of Preservation

There are various degrees of preservation by heating; a few terms have to be identified and understood.

a. Sterilisation. By sterilisation we mean complete destruction of micro-organisms. Because of the resistance of certain bacterial spores to heat, this frequently means a treatment of at least 121° C (250° F) of wet heat for 15 minutes or its equivalent. It also means that every particle of the food must receive this heat treatment. If a can of food is to be sterilised, then immersing it into a 121° C pressure cooker or retort for the 15 minutes will not be sufficient because of relatively slow rate of heat transfer through the food in the can to the most distant point.
b. "Commercially sterile". Term describes the condition that exists in most of canned or bottled products manufactured under Good Manufacturing Practices procedures and methods; these products generally have a shelf-life of two years or more.
c. Pasteurized means a comparatively low order of heat treatment, generally at a temperature below the boiling point of water. The more general objective of pasteurization is to extend product shelf-life from a microbial and enzymatic point of view; this is the objective when fruit or vegetable juices and certain other foods are pasteurized. Pasteurization is frequently combined with another means of preservation - concentration, chemical, acidification, etc.
d. Blanching is a type of pasteurization usually applied to vegetables mainly to inactivate natural food enzymes. Depending on its severity, blanching will also destroy some microorganisms.

Determining Heat Treatment/Thermal Processing Steps

Since heat sufficient to destroy micro-organisms and food enzymes also usually has adverse effects on other properties of foods, in practice the minimum possible heat treatment should be used which can guarantee freedom from pathogens and toxins and give the desired storage life; these aims will determine the choice of heat treatment.

In order to safely preserve foods using heat treatment, the following must be known:

a. What time-temperature combination is required to inactivate the most heat resistant pathogens and spoilage organisms in one particular food?
b. What are the heat penetration characteristics in one particular food, including the can or container of choice if it is packaged?

Preservation processes must provide the heat treatment which will ensure that the remotest particle of food in a batch or within a container will reach a sufficient temperature, for a sufficient time, to inactivate both the most resistant pathogen and the most resistant spoilage organisms if it is to achieve sterility or "commercial sterility", and to inactivate the most heat resistant pathogen if pasteurization for public health purposes is the goal.

Different foods will support growth of different pathogens and different spoilage organisms so the target will vary depending upon the food to be heated.Food acidity/pH value has a tremendous impact on the target in heat preservation/processing. Table given below lists various types of fruit and vegetables and their pH value, together with the heat processing requirements.

Table: Heat processing requirements — dependence on product acidity

Acidity class requirements	pH value processing	Food item	Heat and
Low acid	6.0	Peas, carrots, beets, potatoes, asparagus	High temperature processing 116-121°C (240-250°F)
	5.0	Tomato soup	
Medium acid	4.5	Tomatoes, pears, apricots, peaches	Boiling water processing 100°C (212°F)
Acid	3.7	Sauerkraut, apple,	
High acid	3.0	Pickles	

Sequence of Operations Employed in Heat Preservation of Foods (Fruit and Vegetables, etc.)

In a simplified manner, the main operations employed in heat preservation can be described as follows:

Food preparation	Preparation procedures will vary with the type of food. For fruit, washing, sorting, grading, peeling, cutting to size, pre-cooking and pulping operations may be employed.
Can/receptacle	This may be carried out manually or by using sophisticated filling machinery. The ratio of liquid to solid in the can must be carefully controlled and the can must not be overfilled. A headspace of 6-9 mm depth (6-8% of the container volume) above the level of food in the can is usual. Vacuum production This can be achieved by filling the heated product into the can, by heating the can and contents after filling, by evacuating the headspace gas in a vacuum chamber, or by injecting superheated steam into the headspace. In each case the can end is seamed on immediately afterwards.
Thermal processing	The filled sealed can must be heated to a high temperature for a sufficient length of time to ensure the destruction of spoilage micro-organisms. This is usually carried out in an autoclave or retort, in an environment of steam under pressure.
Cooling	The processed cans must be cooled in chlorinated water to a temperature of 37°C. At this temperature the heat remaining is sufficient to allow the

water droplets on the can to evaporate before labelling and packing. Labelling and packing Labels are applied to the can body, and the cans are then packed into cases.

In principle, all these operations can also be carried out at the farm/ community level using the appropriate small scale equipment, preferably only glass jars (e.g. no metal cans).

Technological principles of pasteurization

Physical and chemical factors which influence pasteurization process are the following:

a. Temperature and time;
b. Acidity of the products;
c. Air remaining in containers.

Pasteurization processes. In pasteurising certain acid juices for example, there are two categories of processes:

a. Low pasteurization where pasteurization time is in the order of minutes and related to the temperature used; two typical temperature/ time combinations are as following:
   63° C to 65° C over 30 minutes
   or
   75° C over 8 to 10 minutes.

Pasteurization temperature and time will vary according to:

- Nature of product; initial degree of contamination;
- Pasteurized product storage conditions and shelf life required.

In this first category of pasteurization processes it is possible to define three phases:

- Heating to a fixed temperature;
- Maintaining this temperature over the established time period (= pasteurization time);
- Cooling the pasteurized products: natural (slow) or forced cooling.

b. Rapid, high or flash pasteurization is characterized by a pasteurization time in the order of seconds and temperatures of about 85° to 90° C or more, depending on holding time. Typical temperature/time combinations are as follows:
   88° C (190° F) for 1 minute;
   100° C for 12 seconds;
   121°C for 2 seconds.

While bacterial destruction is very nearly equivalent in low and in high pasteurization processes, the 121° C/2 seconds treatment give the best quality products in respect of flavour and vitamin retention. Such short holding times, however, require special equipment which is more difficult to design and generally is more expensive than the 63-65 ° C/30 minutes type of processing equipment.

In flash pasteurization the product is heated up rapidly to pasteurization temperature, maintained at this temperature for the required time, then rapidly cooled down to the temperature for filling, which will be performed in aseptic conditions in sterile receptacles. Taking into account the short time and rapid performance of this operation, flash pasteurization can only be achieved in continuous process, using heat exchangers.Industrial applications of pasteurization process are mainly used as a means of preservation for fruits and vegetable juices and specially for tomato juice.

Thermopenetration

The thermopenetration problem is extremely important, especially in the case of the pasteurization of products packed in glass containers because it is the determining factor for the success of the whole operation.

During pasteurization it is necessary that a sufficient heat quantity is transferred through the receptacle walls; this is in order that the product temperature rises sufficiently to be lethal to micro-organisms throughout the product mass. The most suitable and practical method to speed up thermopenetration is the movement of receptacles during the pasteurization process. Rapid rotation of receptacles around their axis is an efficient means to accelerate heat transfer, because this has the effect, among others of rapidly mixing the contents. The critical speed of for this movement is generally about 70 rotations per minute (RPM). This enables a more uniform heating of products, reducing heating time and organoleptic degradation.

Heating may precede or follow packaging. These principles of different temperature time combinations very largely determine the design parametres for heat preservation equipment and commercial practices.

The food processor will employ no less than that heat treatment which gives the necessary degree of micro-organism destruction. This is further ensured by periodic inspection by local sanitary authorities or by the importing countries sanitary services. However, the food processor also will want to use the mildest effective heat treatment to ensure highest food quality.It is convenient to separate heat preservation practices into two broad categories: one involves heating of foods in their final containers, the other employs heat prior to packaging. The latter category includes methods that are inherently less damaging to food quality, where the food can be readily subdivided (such as liquids) for rapid heat exchange. However, these methods then require packaging under aseptic or nearly aseptic conditions to prevent or at least minimise recontamination.On the other hand, heating within the package frequently is less costly and produces quite acceptable quality with the majority of foods and most of our present canned food supply is heated in the package. In practice, therefore, most of the canned food produced locally in developing countries should be heated within the package.

FOOD IRRADIATION

Food irradiation is one of the food processing technologies available to the food industry to control organisms that cause food-borne diseases and to reduce food losses due to spoilage and deterioration. Food irradiation technology offers some advantages over conventional processes. Each application should be evaluated on its own merit as to whether irradiation provides a technical and economical solution that is better than traditional processing methods.

Table: Possible Causes of Spoilage (Real or Aparent) in Canned Goods

Type of food: Acid and high acid foods (canned fruits)	
Condition of can	**Action to be taken to identify cause**
	Insufficient vacuum or headspace Check vacuum and headspace in relation to storage temperature and altitude "Springer" or "flipper"Cool can to 15°C and check if still domed. Check can for denting, if possible measure headspace volume change brought about by dents, by comparing can volume with volume of a sound can
Hydrogen swell	Check degree of detinning in can especially at the liquid level. Look for scratches or pinholes in lacquer or tin coating. Check if can is still domed on cooling to 15°C. "Hard" or "soft swell" Leaker spoilage. Check can for gross seam faults, perforation due to corrosion or damage to seams. Examine contents for signs of spoilage and can interior for detinning at air/product interface.

Applications

For products where irradiation is permitted, commercial applications depend on a number of factors including the demand for the benefits provided, competitiveness with alternative processes and the willingness of consumers to buy irradiated food products. There are a number of applications of food irradiation. For each application it is important to determine the optimum dosage range required to achieve the desired effect. Too high a dosage can produce undesirable changes in texture, colour and taste of foods.

Shelf-life Extension

Irradiation can extend the shelf-life of foods in a number of ways. By reducing the number of spoilage organisms (bacteria, mould, fungi), irradiation can lengthen the shelf life of fruits and vegetables.

Since ionising radiation interferes with cell division, it can be used as an alternative to chemicals to inhibit sprouting and thereby extend the shelf life of potatoes, onions and garlic. Exposure of fruits and vegetables to ionising radiation slows their rate of ripening. Strawberries, for example, have been found to be suitable for irradiation. Their shelf-life can be extended three-fold, from 5 to 15 days.

Disinfestation

Ionising radiation can also be used as an alternative to chemical fumigants for disinfestation of grains, spices, fruits and vegetables. Many countries prohibit the importation of products suspected of being contaminated with live insects to protect the importing country's agricultural base. With the banning of certain chemical fumigants, irradiation has the potential to facilitate the international shipment of food products.

Global Developments

In 1980, an FAD/IAEA/WHO Expert Committee reviewed in detail all the accumulated data on food irradiation from the past 40 years. The Expert Committee concluded that irradiation to an overall dose of 10 kGy (kilograys) presents no toxicological hazard and introduces no special nutritional or microbiological problems, thus establishing the wholesomeness of irradiated foods up to an overall average absorbed dose of 10 kGy.

Data were insufficient to formulate conclusions on applications of food irradiation above 10 kGy. Data on radiation chemistry, nutritional and microbiological aspects of food treated above 10 kGy is currently being compiled.

In 1983, the Codex Alimentarius Commission, an international group that develops global food standards for the FAO and the WHO, incorporated the 1980 Expert Committee's conclusions regarding the wholesomeness of irradiated foods into the Codex General Standard for Irradiated Foods. This proposed international standard was submitted to member countries to accept or to modify according to individual country needs. Currently most countries that allow food irradiation approve its use on a case-by-case basis.

The Codex Alimentarius Commission has also adopted a Recommended International Code of Practice for the Operation of Radiation Facilities for the Treatment of Foods. It is intended to serve as a guide for irradiator operators and government regulators.

International Trade

More than 30 countries have given clearances for the use of food irradiation to process some 40 food items and approximately 30 facilities world-wide treat food by irradiation processing. Approvals for additional items are being considered in many countries and many food irradiation facilities are being planned. It was anticipated in 1988 that by 1990 there could be approximately 50 commercial/demonstration irradiators in 25 countries. Table given below shows commercial applications of food irradiation to fruits and vegetables by country.

Table: International Commercial Applications of Radiation for Fruit and Vegetables

Country	Location (application date)	Food Commodity
Argentina	Buenos Aires (1986)	Spinach

Belgium	Fleurus (1981)	Dehydrated vegetables
Brazil	Sao Paulo (1985)	Dehydrated vegetables
Chile	Santiago (1983)	Dehydrated vegetables onions, potatoes
China	Shanghai (1985	Potatoes
Cuba	Havana (1987)	Potatoes, onions
German Dem. Rep	Weideroda (1983)	Onions, garlic
	Spickendorf (1986)	Onions
Japan	Hokkaido (1973)	Potatoes
Korea	Seoul (1985	Garlic powder
Netherlands	Ede (1978)	Dehydrated vegetables
South Africa	Johannesburg (1981)	Dehydrated vegetables
	Tzaneen (1981)	Fruits, onions, potatoes
Thailand	Bangkok (1971)	Onions

6

Sustainable Agriculture, Food Security and Climate Change

The agricultural sector is a driving force in the gas emissions and land use effects thought to cause climate change. In addition to being a significant user of land and consumer of fossil fuel, agriculture contributes directly to greenhouse gas emissions through practices such as rice production and the raising of livestock; according to the Intergovernmental Panel on Climate Change, the three main causes of the increase in greenhouse gases observed over the past 250 years have been fossil fuels, land use, and agriculture.

LAND USE

Agriculture contributes to greenhouse gas increases through land use in four main ways:

- CO_2 releases linked to deforestation
- Methane releases from rice cultivation
- Methane releases from enteric fermentation in cattle
- Nitrous oxide releases from fertilizer application.

Together, these agricultural processes comprise 54% of methane emissions, roughly 80% of nitrous oxide emissions, and virtually all carbon dioxide emissions tied to land use. The planet's major changes to land cover since 1750 have resulted from deforestation in temperate regions: when forests and woodlands are cleared to make room for fields and pastures, the albedo of the affected area increases, which can result in either warming or cooling effects, depending on local conditions. Deforestation also affects regional carbon reuptake, which can result in increased concentrations of CO_2, the dominant greenhouse gas. Land-clearing methods such as slash and burn compound these effects by burning biomatter, which directly releases greenhouse gases and particulate matter such as soot into the air.

LIVESTOCK

Livestock and livestock-related activities such as deforestation and increasingly fuel-intensive farming practices are responsible for over 18% of human-made greenhouse gas emissions, including:

- 9% of global carbon dioxide emissions
- 35-40% of global methane emissions (chiefly due to enteric fermentation and manure)
- 64% of global nitrous oxide emissions (chiefly due to fertilizer use.)

Livestock activities also contribute disproportionately to land-use effects, since crops such as corn and alfalfa are cultivated in order to feed the animals.

Worldwide, livestock production occupies 70% of all land used for agriculture, or 30% of the land surface of the Earth.

AGRICULTURE AND FOOD SECURITY

Pre-Independence India suffered repeated famines, drought and food shortages. But following the Green Revolution in the '60s, yields and foodstocks rose manifold. Now, 30 years later, Indian farmers have realised the follies of their tryst with intensive agriculture. Despite 70 per cent of the population being engaged in agriculture and allied activities, declining foodgrain production and access to food remain the two biggest problems confronting the country. Liberalisation has made things worse: commercial crops are eating into the fertile land tracts meant for essential foodgrains. And six years after the World Trade Organisation came into existence, the anticipated gains for India from the trade liberalisation process in agriculture are practically zero.

Agriculture Minister C Subramaniam sat huddled with his senior staff. There was tension in the air. Depending on imported foodgrains coming from 12,000 miles away, particularly when it came in driblets following US President Lyndon Johnson's directive to "teach India a lesson", was humiliating enough. What worried him was that food stocks had reached such a precarious low that they were only sufficient for another two weeks. Still worse, there was nothing in transit. Recalls Subramaniam, "As a last resort, I told my officials and experts to identify the nearest food-carrying ships on the ocean throughout the world. I said we would identify those carrying wheat to other countries and appeal to the US President to divert them to India if other countries could wait for another six to eight weeks."

This was in 1966-67, the critical year of drought when India imported 11 million tonnes of foodgrain. A year earlier, India had imported 10 million tonnes of foodgrain. Around this time, the Paddock brothers, often referred to as 'prophets of doom', concluded in their book, Famine 1975, that by the mid-'70s at least half of India would be led to a slaughter house. Considering that India had a long history of hunger and famine, and the subsequent failure of the government's effort to boost agricultural production in the post-Independence era, not many disagreed with the conclusive analysis drawn by William and Paul Paddock.

After all, for a country literally surviving on a 'ship-to-mouth' existence, there was little hope. Soon after Independence, India had begun to seek food aid and in fact emerged as the biggest food importer of this century. In 1951, India got ad hoc assistance of 2 million tonnes of foodgrain to tide over the crisis arising from severe drought in several parts of the country. This was the first major food security on a significant scale after Independence.

Before 1947, Indian history was replete with famine, drought and food shortages. Between 1770 and 1880, as many as 27 food scarcities and famines were recorded. At least 20 million lives were lost in India in about 20 famines that had struck since 1850.

Much of this loss was because of misplaced colonial policies which aimed to derive maximum economic gain at the cost of human suffering and misery. After the British had created a transport infrastructure in the first half of the 19th century, they began encouraging farmers to grow crops that could be exported. The boom in export trade accompanied by rising prices made farmers shift to cash crops like cotton, indigo, poppy and sugarcane. The area under foodgrain subsequently shrank. In other words, efforts to improve agriculture in colonial India were directly linked to the needs of British industries.

In the first half of the 20th century, India's foreign trade showed the same colonial pattern with exports mainly comprising foodgrain and cotton, jute, oilseeds, opium and indigo. Against this, imports consisted mainly of consumer goods. Meanwhile, the cultivator's preference for commercial crops continued.

As is happening now, the trend during the pre-Independence years showed that the best lands available were used for the cultivation of commercial crops. The cultivation of foodgrains was relegated, to some extent, to inferior lands, the yield of which is much less even in normal years. The result was a steady decline in foodgrain output till the Second World War. Per capita food availability from domestic production too had declined. As B M Bhatia observes in The Famines of India: "India kept the precarious balance between increasing population and declining food production partly by turning from an exporter into a net importer of foodgrains but largely by the easy, though dangerous, expedient of pushing down the already low per capita cereal consumption to still lower levels."

The big gap between minimum requirement and supply continued throughout the post-War period. Till the 1960s, Indian agriculture was not sufficiently able to meet domestic requirements and the country had to rely on foodgrain imports.

SOWING THE SEEDS OF THE GREEN REVOLUTION

Realising the disastrous consequences of the growing gap between the rates of growth of population and food production, a vigorous 'grow more food' campaign was launched. The key goal of economic planning shifted to attaining self-sufficiency. The First Five-Year Plan, following food imports in 1951, gave maximum importance to the growth of agriculture. Among the measures initiated in the '50s to stimulate foodgrain production were land reforms, irrigation, fertiliser production, strengthening of research and the organisation of a national farm extension service. Consequently, increased harvests in the mid-1950s relieved the economy.

It was around this time that the government adopted the US Land Grant model of agricultural research. Agricultural universities, beginning with the Gobind Ballabh Pant University of Agriculture and Technology in Pantnagar in Uttar Pradesh, were gradually set up in all the states. The Land Grant model entailed linking research with extension, and taking research from the laboratories to the farms through an extension network linked with the university. In 1961, the government launched the Intensive Agriculture District Programme (IADP) to enhance productivity in the irrigated areas. Although the aim was to introduce improved seed along with a package of agronomic practices to take advantage of the available irrigation, the programme did not come up to expectations for want of high-yielding varieties of wheat.

The tall Indian wheat varieties did not provide an economic response to the application of chemical fertilisers and irrigation water, which agricultural scientists were desperately looking for. These had a tendency to fall down or lodge when fertilisers were applied, as a result of which the yields stagnated at less than one tonne per hectare. Even though breeding for improved non-lodging wheat varieties was initiated in the '50s, it did not meet with much success.

The missing link was provided by Dr. Norman Borlaug, who was then working with CIMMYT in Mexico. A few dwarf wheat strains of spring wheat that he sent to the Indian Agricultural Research Institute in New Delhi, looked impressive. In the next two to three years, a wide range of dwarf material was tested under the All India Wheat Research Project. In August 1964, the then Professor and Head of Department of Plant Breeding and Genetics at the IARI, Dr. M S Swaminathan, who later emerged as the architect of India's Green Revolution, proposed the launching of a National Demonstration Programme in farmers' fields. This would verify the results obtained in research plots and introduce farmers to new opportunities opened up by semi-dwarf varieties for considerably improving the productivity.

Says Swaminathan, "When small farmers, who with the help of scientists organised the National Demonstration Programme, harvested over 5 tonnes of wheat per hectare, its impact on the minds of other farmers was electric. The clamour for seeds began and the area under high-yielding varieties rose from 4 hectares in 1963-64 to over 4 million hectares in 1971-72. A small government programme became a mass movement." Subsequently, the government imported 18,000 tonnes of semi-dwarf wheat from Mexico in 1966 to be distributed in time for immediate sowing by farmers in the northwestern regions.

The seeds of the Green Revolution were, therefore, truly sown in the mid-'60s. What followed next is already part of contemporary history. Wheat production rose to 17 million tonnes in 1968, an increase of 5 million tonnes over the highest of 12 million tonnes harvested in 1964. Indira Gandhi officially recorded the impressive strides in agriculture by releasing a special stamp entitled 'Wheat Revolution' in July 1968. The success of wheat was later replicated in rice. Productivity increases were also recorded in cotton, sugarcane, millets and oilseeds.

MEASURES TO AVOID FAMINE

The famine-avoidance strategy that India launched had all the necessary ingredients for a sustained effort. In 1965, the Agricultural Prices Commission was constituted, with the basic objective of assuring fair prices for farm produce. The Commission, an autonomous body, works out the cost of cultivation for 22 agricultural commodities, and suggests a price to the government that includes profit that should serve as an incentive to grow more. The administered price of wheat, for instance, was raised by 71 per cent between 1972 and 1981. Similarly, the price of rice was jacked up by 81 per cent during the same period. Later, to sustain the tempo of the rice revolution, the relative price structure was moved in its favour. In the early phase of the Green Revolution, the price policy favoured wheat, but after the mid-'70s it tilted in favour of rice. For oilseeds and pulses, the reverse was put into action. Relatively high support prices were announced in the early-'90s to make their cultivation profitable.

Perhaps the most significant aspect of the price support mechanism has been the insulation of farmers against a decline in prices. Wheat production was very high in 1971-72, as a consequence of which prices should have declined by at least 25 per cent. Since the government purchased all the wheat that was offered, farmers were actually paid 7 per cent more. Again, in 1982-83, when wheat production was 14.3 per cent higher than the previous year, the prices should have actually declined by 32.5 per cent. With the government purchasing all the wheat that flowed into the markets, however, the farmers realised 18 per cent more.

The biggest achievement of the Indian food policy, dependent more and more on buffer stocking and operational stock holding, has been the avoidance of famine-like conditions. It was with the basic objective of curbing consumption and ensuring an equitable distribution of available food supplies, especially in the deficit areas and among the poorer strata of society, that the Public Distribution System (PDS) was introduced. PDS supplies were also used for the 'food for work' programme as well as other anti-poverty programmes. At present, PDS covers more than 80 million families and the total foodgrains distributed through a network of 40,000 fair price shops accounts for over one-third of the total trade in foodgrain.

AIMING AT SELF-SUFFICIENCY

The Green Revolution was certainly not a stroke of luck. The resilience was the result of progressive farm policies, determination, sustained efforts of scientists and the vigour of the Indian farmers who readily adopted the improved technology. This is what the Rockefeller Foundation had to say about the magnificent strides taken: "... The speed with which India's farmers and scientists suddenly gave their country an approach to an abundant food supply, has never been duplicated on an equal scale anywhere in the world, including the scientifically sophisticated United States."

Since the early-'70s, the country has avoided famine even during adverse weather conditions, through a carefully designed food security system involving the maintenance of both substantial grain reserves and an extensive PDS. The devastating drought of 1987, rated as the most severe of this century, affected nearly 155 million hectares. In spite of this, no starvation and hunger deaths were reported. This was essentially the result of excellent food management based on foodgrain reserves built largely from homegrown wheat and rice.

From the late-'60s, the rate of growth in food production generally exceeded the rate of growth of population. Despite the continuous growth in population, the per capita availability of foodgrain increased from 157 kg per year in 1955 to 177 kg per year in 1995, an increase of 15.7 per cent. Food production has risen roughly four times since the days of Independence, from 50 million tonnes in 1950-51 to 198 million tonnes in 1996-97. At the same time, the population has risen from 300 million to 960 million. Milk production too has risen to 68 million tonnes, the highest in the world. Food self-sufficiency was achieved in the late-'70s with the maintenance of a sizeable food buffer. Lately, India has begun to export foodgrain. About 5 million tonnes of rice and wheat were exported during the last few years. Skimmed milk powder amounting to about 5,000 tonnes has also been exported.

The increase in foodgrains was, however, disproportionate. The highest rate of growth was achieved in Punjab and Haryana, where, at a compound growth rate of 4.63 per cent, foodgrain production jumped from 7.23 million tonnes in 1964-65 to reach an all-time high of 30.33 million tonnes in 1995-96. Uttar Pradesh recorded a compound growth rate of 3.01 per cent. Production in Maharashtra, Madhya Pradesh, Bihar, Orissa and the northeastern states continues to stagger. Tamil Nadu and Andhra Pradesh, on the other hand, have recorded significant increases in rice yields, with production rising to 9.17 million tonnes and 11.58 million tonnes, respectively. The Green Revolution was aimed at the better-endowed regions. For millions of farmers languishing in the drylands, constituting more than 70 per cent of the cultivable lands, it continues to be a futile struggle. Despite emphasis on dryland farming during the past several decades, the scenario still remains grim. The undulating topography and the irregular rainfall pattern have combined to aggravate the situation. That the drylands produce about 42 per cent of the country's food shows that the future of farming lies in these areas. Nearly 83 per cent of sorghum, 81 per cent of pulses and 90 per cent of oilseeds grown in the country come from these areas. The poor yields and the fluctuations in production are indications of the scant attention drylands have received from policymakers and planners.

The problem of increasing productivity on drylands has serious socio-economic implications. With every passing year, the gap between the farmer's yields in irrigated areas and in the dry farming region is widening. One year of drought is enough to push a farmer into a deep well of poverty for another two to three years. Drought is a recurring phenomenon in arid and semi-arid areas. Fifty years after Independence, life for millions of people somehow surviving in the drylands continues to be worse than before.

DECLINE IN PRODUCTION

Thirty years after the dawn of the Green Revolution, Indian farmers are realising that their love affair with intensive agriculture is on the decline. Despite a bountiful monsoon for the tenth year in a row, harvests are not as plentiful as expected. However, with no let-up in the ever-increasing rise in population, Indian agriculture is once again at the crossroads.

Alarm bells have been ringing for quite some time. For nearly a decade, agricultural production had stagnated. The spectacular yield growth recorded in the post-Green Revolution years in Punjab and Haryana has receded into history. Of the multiple problems confronting agriculture, rapid fragmentation of land holdings is keeping pace with increasing population. In 1976-77, the average size of the holdings was estimated at 2 hectares, while in 1980-81, it came down to 1.8 hectares. Today, it stands at a mere 0.2 hectares. The total number of land holdings in 1981 were around 89 million; today these have crossed 100 million. By the turn of the century, the average land holding will come down to 0.11 hectares.

It is quite obvious that with such small land holdings, Indian agriculture cannot adopt high-tech farm practices. Basking in the afterglow of the Green Revolution, farming and agriculture have ceased to attract serious attention. In 1995-96, foodgrain production slumped to minus 3.60 per cent. In 1997-98, it further declined to minus 3.70 per cent, the worst-ever since the heady days of the Green Revolution. Foodgrain production in the frontline agricultural states of Punjab, Haryana and western Uttar Pradesh, comprising the country's food bowl, has decelerated. The miracle that began with wheat was replicated in rice. This is where the cost-benefit ratio has turned negative. The downward spiral in food production continues through the southern regions of the country. Tamil Nadu, another Green Revolution area, is under tremendous strain from intensive cultivation. In Karnataka, the negative trend in yield levels of all food crops, barring cotton and sugarcane, are all too apparent. Farming in Karnataka can be clearly separated in two distinct classes, the 'creamy layer' of corporate agriculture occupying the fertile and irrigated areas, and the remaining low productive tracts at the mercy of the rains, constituting the tiny and small land holdings. Kerala too, picturesque and verdant with tropical forests, is passing through its worst-ever crisis on the foodgrain front. As if this is not enough, the number of landless in rural areas too has been multiplying over the past few decades, at an estimated rate of two million every year. The rapid expansion in the number of land holdings, with less than a hectare under plough, has meant an additional 6,00,000 farmers every five years or so. The negative terms of trade for agriculture and the declining public sector investments in farming are indices of the sluggishness in this sector.

The Economic Survey for 1996-97 has further sounded a warning. The compound growth rate in foodgrain production at 1.7 per cent between 1991 and 1996, is lower than the annual population growth of 1.9 per cent. The survey adds that even a marginal fall of three to four per cent in foodgrain output can cause the prices of essential commodities to escalate sharply, necessitating government intervention by way of resorting to imports. This happened in 1993 and again in 1996.

The philosophy of agricultural planning is changing. Gone are the days when the nation's emphasis was solely on attaining self-sufficiency in foodgrain production. Gone also are the days when farmers were newly independent India's heroes, revered for their role in keeping hunger and starvation at bay. Today, at a time when food production struggles to keep pace with the burgeoning population growth, farmers are being asked to diversify, produce crops that are suitable for export and to compete in the international market. With the promise of cheap food available off the shelf in the global market, the focus has shifted from agriculture to industry, trade and commerce, from the small and marginal farmers to the agri-processing companies, which alone can bring in investments and add value to produce.

The restructuring of the Indian economy too has put the emphasis on agri-exports. Cultivation of staple food is being replaced by cash crops, tomatoes in place of wheat, durum wheat (for bakery purposes) replacing wheat as a staple diet in Punjab and Haryana, flowers in place of rice, and so on. In the coastal areas, private enterprise is taking away the fish catch, depriving the local communities of a livelihood and their only source of nutrition. In Kerala, vast tracts of forests and paddy fields have been converted into rubber, coffee and coconut plantations. Every year, about 25,000 hectares of good paddy land are being diverted for non-paddy purposes. The structural transformation is not only peculiar to Kerala. It is happening in almost all the states. Commercial crops are eating into the fertile land tracts meant for growing essential foodgrains, thereby making the states rely more heavily on the PDS. The diversion of good agricultural land, which in any case is limited, to commercial farming and even industries, is further restricting the ability to grow enough foodgrain.

ENVIRONMENTAL IMPACT OF AGRICULTURE

The environmental impact of agriculture varies based on the wide variety of agricultural practices employed around the world.

Climate change and agriculture are interrelated processes, both of which take place on a global scale. Global warming is projected to have significant impacts on conditions affecting agriculture, including temperature, precipitation and glacial run-off. These conditions determine the carrying capacity of the biosphere to produce enough food for the human population and domesticated animals.

Rising carbon dioxide levels would also have effects, both detrimental and beneficial, on crop yields. The overall effect of climate change on agriculture will depend on the balance of these effects. Assessment of the effects of global climate changes on agriculture might help to properly anticipate and adapt farming to maximize agricultural production. At the same time, agriculture has been shown to produce significant effects on climate change, primarily through the production and release of greenhouse gases such as carbon dioxide, methane, and nitrous oxide, but also by altering the Earth's land cover, which can change its ability to absorb or reflect heat and light, thus contributing to radiative forcing. Land use change such as deforestation and desertification, together with use of fossil fuels, are the major anthropogenic sources of carbon dioxide; agriculture itself is the major contributor to increasing methane and nitrous oxide concentrations in earth's atmosphere.

INTENSIVE FARMING

Intensive farming alters the environment in many ways. Some of the disadvantages of this method of farming include:

- Limits or destroys the natural habitat of most wildlife, and leads to soil erosion
- Use of fertilizers can alter the biology of rivers and lakes.
- Pesticides generally kill useful insects as well as those that destroy crops
- Generally not sustainable-often results in desertification or, in a worst case scenario, land that is so poisonous and eroded that nothing else will grow
- Requires large amounts of energy input to produce, transport, and apply chemical fertilizers/pesticides
- Use of chemicals on fields creates run-off, excess runs off into rivers and lakes causing pollution
- Use of pesticides have numerous negative health effects in workers

who apply them, people that live nearby the area of application or downstream/downwind from it, and consumers who eat the pesticides which remain on their food

Irrigation

Irrigation can lead to a number of problems:

- Depletion of underground aquifers through overdrafting.
- Ground subsidence.
  - Groundwater recharge-an ecological restoration, mitigation, and remediation technique.
- Underirrigation gives poor soil salinity control which leads to increased soil salinity with consequent build up of toxic salts on soil surface in areas with high evaporation. This requires either leaching to remove these salts and a method of drainage to carry the salts away.
- Overirrigation because of poor distribution uniformity or management wastes water, chemicals, and may lead to water pollution.
- Deep drainage (from over-irrigation) may result in rising water tables which in some instances will lead to problems of irrigation salinity requiring watertable control by some form of subsurface land drainage.
- Irrigation with saline or high-sodium water may damage soil structure owing to the formation of alkaline soil.
- Runoff causing surface water and groundwater-aquifer hydrologic cycle water pollution.
  - Bioretention-an ecological restoration, mitigation, and remediation technique.

Pollutants

A wide range of agricultural chemicals are used and some become pollutants through use, misuse, or ignorance.

- Pesticide drift
  - soil contamination
  - groundwater and water pollution
  - air pollution *spray drift*
- Pesticides, especially those based on organochloride
- Pesticide residue in foods
- Pesticide toxicity to bees
  - List of crop plants pollinated by bees
  - Pollination management
- Bioremediation.

Soil Degradation

- soil contamination

- erosion
  - sedimentation.

Waste

Plasticulture, the use of plastic materials in agriculture, raises problems around how to carry out the recycling of agricultural plastics.

Issues by Region

- Hedgerow removal in the United Kingdom.
- Soil salinisation, especially in Australia.
- Phosphate mining in Nauru
- Methane emissions from livestock in New Zealand.
- Some environmentalists attribute the hypoxic zone in the Gulf of Mexico as being encouraged by nitrogen fertilization of the algae bloom.

SUSTAINABLE AGRICULTURE: CLIMATE CHANGE AND FOOD SECURITY

The exponential population increase in recent decades has increased the practice of agricultural land conversion to meet demand for food which in turn has increased the effects on the environment. The global population is still increasing and will eventually stabilise, as some critics doubt that food production, due to lower yields from global warming, can support the global population. Organic farming is a multifaceted sustainable agriculture set of practices that can have a lower impact on the environment Other specific methods include: permaculture; and biodynamic agriculture which incorporating a spiritual element.

The links between climate change and food security have, to date, largely been explored in relation to impacts on crop productivity and hence, food production. For instance, Gregory *et al.* (1999) summarized experimental findings on wheat and rice that indicated decreased crop duration (and hence yield) of wheat as a consequence of warming and reductions in yields of rice of about 5% $°C^{-1}$ rise above 32 °C. These effects of temperature were considered sufficiently detrimental that they would largely offset any increase in yield as a consequence of increased atmospheric carbon dioxide (CO_2) concentration. Several reviews have further assessed the potential consequences of changes in climate on the growth and yield of crop plants, concluding that the earlier-anticipated benefits of CO_2 fertilization would be largely offset by nutrient limitations, pollutants and further interactions with climatic factors. Similarly, simulations of maize production in Africa and Latin America using climate data from the HadCM2 model to generate characteristic daily weather data for 2055 predicted an overall reduction of 10%. This, and other similar projections, uses a process-based crop production model (such as the crop environment resource synthesis family) to link climate to plant physiological processes. Yield can then be modelled for a uniform crop and upscaled to a larger area normally within some form of geographic information system (GIS). The aggregate result of Jones & Thornton (2003), though, hides considerable variability within and between countries, and, as they point out, also ignores the fact that maize is commonly used as fodder as well as food as part of a complex production system.

The spatial variation in effects of climate change was an explicit component of the study on potential yields of rainfed cereal crops undertaken by Fischer *et al.* (2001). In this analysis, climate predictions in 2080 were obtained from various global circulation models (GCMs) and assessments made based on current populations and socio-economic conditions.

The output demonstrated that cereal producing regions of Canada, and northern Europe and Russia might be expected to increase production as a consequence of the climate changes predicted by GCMs, while many parts of the world would suffer losses including the western edge of the USA prairies, eastern Brazil and western Australia.

Overall, the results of this and subsequent work that included assessments of future populations and alternative future socio-economic conditions, demonstrated that climate change would benefit the developed countries more than the developing countries even if cropping practices evolved to allow more than one rainfed crop per year. Moreover, the anticipated demographic growth and socio-economic development in these developing countries would result in substantial increases in food requirements thereby exacerbating the detrimental effects of climate change.

The spectre of climate change, together with other global environmental changes such as changes in water availability, and land cover, and altered nitrogen availability and cycling (all strongly influenced by human activities), has increased concerns about achieving food security especially for poor people. There is also concern that meeting the global demand for food resulting from higher population and changing dietary preferences will further degrade the environment both through additional destruction of native vegetation and increased intensification of cropped areas. This may, in turn, further undermine the food systems upon which food security is based.

While there has been considerable progress in understanding the sensitivities of crop yield to climate change, assessments of climate change effects on food security remain rather limited. Food security is concerned not only with food availability but also with access to and utilization of, food so that studies, which focus only on crop production provide only a partial assessment of food security–climate change relationships. This is of particular concern as many of the policy-based issues that are relevant to the international and sustainable development communities are posed in the context of food security prospects. Engagement of these policy communities requires a much broader and comprehensive research framework. Recognition that food security depends on robust food systems that encompass issues of availability, access and utilization (not merely production alone), and consequently that the nature of key research issues changes as questions more related to food security are formulated, led the international global change research community to establish the joint project global environmental change and food systems (GECAFS) in 2001. GECAFS is an interdisciplinary research programme that aims to improve understanding of the relationship between food systems and the earth system to deliver science-based tools for analysing socio-economic and environmental consequences of strategies seeking to adapt to climate and other environmental changes. The purposes of this chapter are to: (i) outline the notion of food systems, their relation to food security, and links to climate change; (ii) explore the vulnerability of food systems to environmental change; (iii) examine some prospects for adaptation of food systems in response to climate change and (iv) highlight some of the environmental feedbacks resulting from modifications to food systems.

FOOD SECURITY, FOOD SYSTEMS AND THE LINK TO CLIMATE

There are several definitions of what constitutes food systems each formulated in relation to a specific range of issues. For GECAFSs purpose, food systems are defined as a set of dynamic interactions between and within the biogeophysical and human environments which result in the production, processing, distribution, preparation and consumption of food. They encompass components of: (i) food availability (with elements related to production, distribution and exchange); (ii) food access (with elements related to affordability, allocation and preference) and (iii) food utilization. Food systems, then, involve much broader considerations than productivity and production alone. They underpin food security, which is the state achieved when food systems operate such that 'all people, at all times, have physical and economic access to sufficient, safe and nutritious food to meet their dietary needs and food preferences for an active and healthy life' (FAO 1996). Food security is diminished when food systems are stressed. This can be caused by a range of factors in addition to climate and other environmental changes (*e.g.* conflict, changes in international trade agreements and policies, HIV/AIDS) and may be particularly severe when these factors act in combination. Access to culturally acceptable food by individuals and communities, or means for its procurement, is increasingly being elaborated into human rights legislation.

Food systems may be simple, as in the case of a subsistence farmer who produces, processes and consumes food on farm. However, there are comparatively few individuals or households in the world that are totally self-reliant for food throughout a year, and in almost all cases there is an element of bartering, exchange, or the cash economy to bring food into the household. In many places, the food system has changed radically in the last century and continues to become increasingly complex. The intensification of agricultural production, since the 1940s has been accompanied by profound changes in the organization of food systems around the world including changes in distribution, marketing, affordability and preferences for particular food items.

These changes are especially obvious in the USA and Europe, where market globalization has occurred with global sourcing of products by retailers direct from producers in the case of fresh fruit and vegetables and from a few, large manufacturers for other food products. Such changes have also shifted economic and political power from farmers to retailers, from national legislative bodies to regional and global organizations, and from the state to multinational corporations. For example, in a study of selected OECD countries in Europe, Grievink (2003) determined that the food chain has some 160 million consumers of whom about 3.2 million are also farmers or food producers, but the link between these groups is increasingly determined by the small number of food processors/manufacturers (about 90 000) and the even smaller number of buyers (about 100) for the supermarket chains.

Components of the food chain in selected OECD countries of Europe indicating the two inverted pyramid structure that relates farmers to consumers. (From Grievink 2003.) Food systems around the world are changing very rapidly as urbanization and globalization proceed apace. The urbanization of many predominantly rural countries in the last three decades has been accompanied by the rapid growth of supermarkets in many, often accompanied by foreign investment by global retail chains. However, even in poor countries such as Kenya, where per capita gross domestic product (GDP) was less than \$400 yr^{-1} in 2002, supermarkets have grown from a tiny niche market in 1997 to be greater than 20% of urban food retailing today. This growth has been a consequence of three major factors. First, rapid urbanization has seen the proportion of the population based in urban areas grow from 0.13 in 1975 to 0.36 in 2000 with an expectation of greater than 0.5 in 2013; second, trade and domestic market liberalization, since 1993 has seen the removal of import controls and the deregulation of prices and third, movement towards price-based competition between the indigenous chains. Success within Kenya is now spreading to other East African countries with important effects on the market conditions faced by farmers including the decline of traditional wholesalers (and the smallholder producers from whom they buy) and the increase in direct purchases from larger farms.

These marked changes in access and utilization of food around the world provide a context in which to evaluate the likely effects of climate and other environmental changes on crop production and food security. Not all food systems or parts of food systems are equally vulnerable to environmental changes because the capacity to cope with existing variability in bio-physical and socio-economic systems, and the ability of humans to perceive environmental changes and to adapt food systems, differs. Human vulnerability includes both the likelihood of exposure to stresses as well as the capacity to cope with such stresses (*i.e.* sensitivity). Vulnerability and poverty are often inter-related because both the likelihood of exposure to stresses is greater among the poor and because a large proportion of their resources are spent either purchasing or producing food, thereby reducing their capacity to cope with perturbations.

Moreover, pursuit of food security frequently involves trade-offs with expenditure on health and education reducing still further the ability to improve longer-term living conditions or resilience to stress and shock. Food insecurity is experienced at a range of spatial scales from individual households to regions, as well as a range of time-scales, and reductions of vulnerability at one scale do not automatically flow to the next scale. For example, regional policy decisions do not always convert to successful local implementation especially if distribution services are inadequate, or food preferences are ignored. Climate variation is one of several interacting factors that affect food security. For example, in studies of household food security in southern Africa, climate/environment was only one of some 33 drivers mentioned as important by householders. The mix of drivers varied across the region but in all communities many interacting factors resulted in vulnerability to food shortages. Overall, however, climate/environment was one of the seven factors influencing food security that were frequently cited, because of its role both as an ongoing issue (57% of cases, where it was mentioned) and as a 'shock' (43%).

The impacts of sudden shocks such as drought are felt, then, on top of ongoing long-term stresses, and the low ability to cope with such shocks and to mitigate long-term stresses means that the employment of coping strategies that might be available to others, is at a too high cost or, simply, unavailable. Typically, reliance on purchased food increases in drought years due to losses in food production leading to an increase in poverty due to the synergistic action of other drivers such as rising food prices and unemployment. Scholes & Biggs (2004) record that the food security crisis in southern Africa in 2002–2003 was not simply a result of drought alone and, indeed, climatic stress was not as severe as in previous crises.

Maize production during the preceding growing season was only 5.5% less than the previous five year average so food stocks at the start of the climatic shock were not unusually low. It was concluded that the crisis was indicative of entrenched vulnerability resulting from a range of regional and global political and economic factors including high food prices, legacies of structural adjustment, government policies, conflict and war, policies on genetically modified foods, and poor responses to the HIV/AIDS pandemic. The key here was that conditions, which weakened food systems in the region were already in place. The moderate climatic shock intensified food insecurity and the long-term vulnerability of the region.

The seven most frequently cited drivers in 49 studies of household-level food insecurity in southern Africa. The numbers in the arrows indicate the number of citations, as a percentage of 555 citations of 33 possible drivers. The drivers shaded in grey were noted as being chronic, while those in white indicate drivers experienced mainly as 'shocks'. The shaded arrows indicate drivers that acted primarily via reductions in food production, while the white arrows indicate those which acted by restricting access to food.

QUANTIFYING THE IMPACTS ON FOOD SECURITY

A number of studies have recently quantified the impacts of climate change on food security. In terms of quantifying agronomic yield change projections, these studies are either based on the AEZ tools developed by the IIASA analysis or the Decision Support System for Agrotechnology Transfer suite of crop models; all use the IIASA-Basic Linked System (BLS) economic model for assessing economic impacts. These tools, with some modifications relating to how crop yield changes are simulated, have also been used by others to undertake similar assessments and provide sensitivity analyses across a range of SRES and general circulation model (GCM) projections.

Many other simulations have also examined the effects of climate change with and without adaptation (induced technological progress, domestic policy change, international trade liberalization, etc.), with and without mitigation (*e.g.*, CO_2 stabilization, variants for temperature, rainfall change and distribution) or provide impact assessments for different speeds of climate change.

The quantitative results for food security, trying to illuminate some of the differences and extract the main messages that emerge from the various studies. Unless indicated, all simulation results discussed below include the combined effects of climate change and elevated CO_2 on crops. The key messages can be summarized as follows:

First, it is very likely that climate change is likely to increase the number of people at risk of hunger compared with reference scenarios with no climate change; the exact impacts will, however, strongly depend on the projected socio-economic developments. For instance, it is estimated that climate change would increase the number of undernourished in 2080 by 5–26%, compared with no climate change or by between 5 million and 10 million (B1 SRES) and 120 million to 170 million people (A2 SRES), with within-SRES ranges depending on GCM climate projections. Using only one GCM scenario, others projected small reductions by 2080, *i.e.*, –5%, or –10 million (B1) to –30 million (A2) people, and slight increases of +13%–26%, or H–10 million (B2) to 30 million (A1) people.

Second, it is likely that the magnitude of these climate impacts will be small compared with the impact of socio-economic development. The limitations of socio-economic forecasts, these studies suggest that robust economic growth and a decline in population growth projected for the 21st century will, in all but one scenario (SRES A2), significantly reduce the number of people at risk of hunger in 2080. At any rate, the prevalence of undernourishment will decline as all scenarios assume that world population will continue to grow to 2080, albeit at lower rates. Compared with FAO estimates of 820 million undernourished in developing countries today, several studies estimate reductions of >75% by 2080, or H–560 million to 700 million people, projecting 100 million to 240 million undernourished by 2080 (A1, B1, and B2). As mentioned, the only exception is scenario A2, where the number of the hungry is forecast to decrease only slightly by 2080; but the higher population growth rates in A2 compared with other scenarios mean that the prevalence of undernourishment will decline drastically. However, these analyses also confirm that the progress will be unevenly distributed over the developing world, and more importantly progress will be slow during the first decades of the outlook. With or without climate change, the millennium development goal of halving the prevalence of hunger by 2015 is unlikely to be realized before 2020–2030.

In addition to socio-economic pressures considered within the IPCC SRES scenarios, food production may increasingly compete with bio-energy in coming decades; studies addressing possible consequences for world food supply have only started to surface, providing both positive and negative views. Importantly, none of the major world food models discussed herein have yet considered such competition.

Third, sub-Saharan Africa is likely to surpass Asia as the most food-insecure region. However, this is largely independent of climate change and is mostly the result of the socio-economic development paths assumed for the different developing regions in the SRES scenarios. Throughout most SRES and climate-change scenarios sub-Saharan Africa accounts for 40–50% of global hunger by 2080, compared with H–24% today; in some simulations sub-Saharan Africa even accounts for 70–75% of global undernourishment by 2080. These high estimates have emerged from slower growth variants of the A2 and B2 scenarios; also an A2 variant with slower population growth yields a sharper concentration of hunger in sub-Saharan Africa. For regions other than sub-Saharan Africa, results largely depend on GCM scenarios and therefore are highly uncertain.

Fourth, although there is significant uncertainty on the effects of elevated CO_2 on crop yields, this uncertainty is carried to a much lesser extent on food security. This result emerges from a comparison of climate change simulations with and without CO_2 fertilization effects on crop yields. As can be seen, higher CO_2 fertilization would not greatly affect global projections of hunger. In view of the fact that essentially all SRES worlds are characterized by much higher real incomes, much improved transportation and communication options, and still sufficient global food production, the somewhat smaller supplies will not be able to make a dent in global food security outcomes.

Many studies find that climate change without CO_2 fertilization would reduce the number of undernourished people by 2080 only by some 20 million to 140 million (120 million to 380 million for SRES A1, B1, and B2 without, compared with 100 million to 240 million with CO_2 fertilization effect). The exception again in these studies is SRES A2, under which the assumption of no CO_2 fertilization results in a projected range of 950 million to 1,300 million people undernourished in 2080, compared with 740 million to 850 million projected under climate change, but with CO_2 effects on crops.

Finally, recent research suggests large positive effects of climate stabilization for the agricultural sector. However, as the stabilizing effects of mitigation measures can take several decades to be realized from the moment of implementation, the benefits for crop production may be realized only in the second half of this century. Importantly, even in the presence of robust global long-term benefits, regional and temporal patterns of winners and losers that can be projected with current tools are highly uncertain and depend critically on the underlying GCM projections.

UNCERTAINTIES AND LIMITATIONS

The finding that socio-economic development paths have an important bearing on future food security and that they are likely to top the effects of climate change should not, or at least not only, be interpreted as a probability-based forecast. This is because SRES scenarios offer a range of possible outcomes "without any sense of likelihood". Yet SRES scenarios, like all scenarios, do not overcome the inability to accurately project future changes in economic activity, emissions, and climate.

Second, the existing global assessments of climate change and food security have only been able to focus on the impacts on food availability and access to food, without quantification of the likely important climate change effects on food safety and vulnerability (stability). This means that these assessments neither include potential problems arising from additional impacts due to extreme events such as drought and floods nor do they quantify the potential impacts of changes in the prevalence of food-borne diseases (positive as well as negative) or the interaction of nutrition and health effects through changes in the proliferation of vector-borne diseases such as malaria. On the food availability side, they also exclude the impacts of a possible sea-level rise for agricultural production or those that are associated with possible reductions of marine or freshwater fish production. Third, it is important to note that even in terms of food availability, all current assessments of world food supply have focused only on the impacts of mean climate change, *i.e.*, they have not considered the possibility of significant shifts in the frequency of extreme events on regional production potential, nor have they considered scenarios of abrupt climate or socio-economic change; any of these scenario variants is likely to increase the already negative projected impacts of climate change on world food supply. Models that take into account the specific biophysical, technological, and market responses necessary to simulate realistic adaptation to such events are not yet available.

Fourth, this review finds that recent global assessments of climate change and food security rest essentially on a single modeling framework, the IIASA system, which combines the FAO/IIASA AEZ model with various GCM models and the IIASA BLS system, or on close variants of the IIASA system. This has important implications for uncertainty, given that the robustness of all these assessments strongly depends on the performance of the underlying models. There is, therefore, a clear need for continued and enhanced validation efforts of both the agro-climatological and food trade tools developed at IIASA and widely used in the literature.

Finally, we note that assessments that do not only provide scenarios, but also attach probabilities for particular outcomes to come true could provide an important element for improved or at least better-informed policy decisions. A number of possibilities are offered to address the related modeling challenges. One option would be to produce such estimates with probability-based estimates of the (key) model parameters. Alternatively, the various scenarios could be constructed so that they reflect expert judgements on a particular issue. It would be desirable to attach probabilities to existing scenarios. Information on how likely the suggested outcomes are would contribute greatly to their usefulness for policy makers and help justify (or otherwise) policy measures to adapt to or mitigate the impacts of climate change on food security. Climate change will affect all four dimensions of food security, namely food availability (*i.e.*, production and trade), access to food, stability of food supplies, and food utilization. The importance of the various dimensions and the overall impact of climate change on food security will differ across regions and over time and, most importantly, will depend on the overall socio-economic status that a country has accomplished as the effects of climate change set in.

Essentially all quantitative assessments show that climate change will adversely affect food security. Climate change will increase the dependency of developing countries on imports and accentuate existing focus of food insecurity on sub-Saharan Africa and to a lesser extent on South Asia. Within the developing world, the adverse impacts of climate change will fall disproportionately on the poor. Many quantitative assessments also show that the socio-economic environment in which climate change is likely to evolve is more important than the impacts that can be expected from the biophysical changes of climate change.

Less is known about the role of climate change for food stability and utilization, at least in quantitative terms. However, it is likely that differences in socio-economic development paths will also be the crucial determinant for food utilization in the long run and that they will be decisive for the ability to cope with problems of food instability, be they climate-related or caused by other factors.

Finally, all quantitative assessments we reviewed show that the first decades of the 21st century are expected to see low impacts of climate change, but also lower overall incomes and still a higher dependence on agriculture. During these first decades, the biophysical changes as such will be less pronounced but climate change will affect those particularly adversely that are still more dependent on agriculture and have lower overall incomes to cope with the impacts of climate change. By contrast, the second half of the century is expected to bring more severe biophysical impacts but also a greater ability to cope with them. The underlying assumption is that the general transition in the income formation away from agriculture towards nonagriculture will be successful.

How strong the impacts of climate change will be felt over all decades will crucially depend on the future policy environment for the poor. Freer trade can help to improve access to international supplies; investments in transportation and communication infrastructure will help provide secure and timely local deliveries; irrigation, a promotion of sustainable agricultural practices, and continued technological progress can play a crucial role in providing steady local and international supplies under climate change.

CLIMATE CHANGE AFFECTS FOOD SECURITY AND AGRICULURE

Climate change affects agriculture and food production in complex ways. It affects food production directly through changes in agro-ecological conditions and indirectly by affecting growth and distribution of incomes, and thus demand for agricultural produce. Impacts have been quantified in numerous studies and under various sets of assumptions. A selection of these results is presented in *Quantifying the Impacts on Food Security*. Here it is useful to summarize the main alterations in the agro-ecological environment that are associated with climate change.

Changes in temperature and precipitation associated with continued emissions of greenhouse gases will bring changes in land suitability and crop yields. In particular, the Intergovernmental Panel on Climate Change (IPCC) considers four families of socio-economic development and associated emission scenarios, known as Special Report on Emissions Scenarios (SRES) A2, B2, A1, and B1. Of relevance to this review, of the SRES scenarios, A1, the "business-as-usual scenario," corresponds to the highest emissions, and B1 corresponds to the lowest. The other scenarios are intermediate between these two. Importantly for agriculture and world food supply, SRES A2 assumes the highest projected population growth of the four (United Nations high projection) and is thus associated to the highest food demand. Depending on the SRES emission scenario and climate models considered, global mean surface temperature is projected to rise in a range from 1.8°C (with a range from 1.1°C to 2.9°C for SRES B1) to 4.0°C (with a range from 2.4°C to 6.4°C for A1) by 2100. In temperate latitudes, higher temperatures are expected to bring predominantly benefits to agriculture: the areas potentially suitable for cropping will expand, the length of the growing period will increase, and crop yields may rise.

A moderate incremental warming in some humid and temperate grasslands may increase pasture productivity and reduce the need for housing and for compound feed. These gains have to be set against an increased frequency of extreme events, for instance, heat waves and droughts in the Mediterranean region or increased heavy precipitation events and flooding in temperate regions, including the possibility of increased coastal storms; they also have to be set against the fact that semiarid and arid pastures are likely to see reduced livestock productivity and increased livestock mortality.

In drier areas, climate models predict increased evapotranspiration and lower soil moisture levels. As a result, some cultivated areas may become unsuitable for cropping and some tropical grassland may become increasingly arid. Temperature rise will also expand the range of many agricultural pests and increase the ability of pest populations to survive the winter and attack spring crops.

Another important change for agriculture is the increase in atmospheric carbon dioxide (CO_2) concentrations. Depending on the SRES emission scenario, the atmospheric CO_2 concentration is projected to increase from H"379 ppm today to >550 ppm by 2100 in SRES B1 to >800 ppm in SRES A1FI. Higher CO_2 concentrations will have a positive effect on many crops, enhancing biomass accumulation and final yield. However, the magnitude of this effect is less clear, with important differences depending on management type (*e.g.*, irrigation and fertilization regimes) and crop type. Experimental yield response to elevated CO_2 show that under optimal growth conditions, crop yields increase at 550 ppm CO_2 in the range of 10% to 20% for C_3 crops (such as wheat, rice, and soybean), and only 0–10% for C_4 crops such as maize and sorghum. Yet the nutritional quality of agricultural produce may not increase in line with higher yields. Some cereal and forage crops, for example, show lower protein concentrations under elevated CO_2 conditions.

Finally, a number of recent studies have estimated the likely changes in land suitability, potential yields, and agricultural production on the current suite of crops and cultivars available today. Therefore, these estimates implicitly include adaptation using available management techniques and crops, but excluding new cultivars from breeding or biotechnology. These studies are in essence based on the FAO/International Institute for Applied Systems Analysis (IIASA) agro-ecological zone (AEZ) methodology. For instance, pioneering work in ref. 9 suggested that total land and total prime land would remain virtually unchanged at the current levels of 2,600 million and 2,000 million hectares (ha), respectively. The same study also showed, however, more pronounced regional shifts, with a considerable increase in suitable cropland at higher latitudes (developed countries +160 million ha) and a corresponding decline of potential cropland at lower latitudes (developing countries "110 million ha). An even more pronounced shift within the quality of cropland is predicted in developing countries. The net decline of 110 million ha is the result of a massive decline in agricultural prime land of H"135 million ha, which is offset by an increase in moderately suitable land of >20 million ha. This quality shift is also reflected in the shift in land suitable for multiple cropping. In sub-Saharan Africa alone, land for double cropping would decline by between 10 million and 20 million ha, and land suitable for triple copping would decline by 5 million to 10 million ha. At a regional level, similar approaches indicate that under climate change, the biggest losses in suitable cropland are likely to be in Africa, whereas the largest expansion of suitable cropland is in the Russian Federation and Central Asia.

IMPACTS ON THE STABILITY OF FOOD SUPPLIES

Global and regional weather conditions are also expected to become more variable than at present, with increases in the frequency and severity of extreme events such as cyclones, floods, hailstorms, and droughts. By bringing greater fluctuations in crop yields and local food supplies and higher risks of landslides and erosion damage, they can adversely affect the stability of food supplies and thus food security. Neither climate change nor short-term climate variability and associated adaptation are new phenomena in agriculture, of course. As shown, for instance, in ref. 9, some important agricultural areas of the world like the Midwest of the United States, the northeast of Argentina, southern Africa, or southeast Australia have traditionally experienced higher climate variability than other regions such as central Africa or Europe. They also show that the extent of short-term fluctuations has changed over longer periods of time. In developed countries, for instance, short-term climate variability increased from 1931 to 1960 as compared with 1901 to 1930, but decreased strongly in the period from 1961 to 1990. What is new, however, is the fact that the areas subject to high climate variability are likely to expand, whereas the extent of short-term climate variability is likely to increase across all regions. Furthermore, the rates and levels of projected warming may exceed in some regions the historical experience.

If climate fluctuations become more pronounced and more widespread, droughts and floods, the dominant causes of short-term fluctuations in food production in semiarid and subhumid areas, will become more severe and more frequent. In semiarid areas, droughts can dramatically reduce crop yields and livestock numbers and productivity. Again, most of this land is in sub-Saharan Africa and parts of South Asia, meaning that the poorest regions with the highest level of chronic undernourishment will also be exposed to the highest degree of instability in food production. How strongly these impacts will be felt will crucially depend on whether such fluctuations can be countered by investments in irrigation, better storage facilities, or higher food imports. In addition, a policy environment that fosters freer trade and promotes investments in transportation, communications, and irrigation infrastructure can help address these challenges early on.

IMPACTS OF CLIMATE CHANGE ON FOOD UTILIZATION

Climate change will also affect the ability of individuals to use food effectively by altering the conditions for food safety and changing the disease pressure from vector, water, and food-borne diseases. The IPPC Working Group II provides a detailed account of the health impacts of climate change of its fourth assessment report. It examines how the various forms of diseases, including vector-borne diseases such as malaria, are likely to spread or recede with climate change. This article focuses on a narrow selection of diseases that affect food safety directly, *i.e.*, food and water-borne diseases.

The main concern about climate change and food security is that changing climatic conditions can initiate a vicious circle where infectious disease causes or compounds hunger, which, in turn, makes the affected populations more susceptible to infectious disease. The result can be a substantial decline in labour productivity and an increase in poverty and even mortality. Essentially all manifestations of climate change, be they drought, higher temperatures, or heavy rainfalls have an impact on the disease pressure, and there is growing evidence that these changes affect food safety and food security.

The recent IPCC report also emphasizes that increases in daily temperatures will increase the frequency of food poisoning, particularly in temperate regions. Warmer seas may contribute to increased cases of human shellfish and reef-fish poisoning (ciguatera) in tropical regions and a poleward expansion of the disease. However, there is little new evidence that climate change significantly alters the prevalence of these diseases. Several studies have confirmed and quantified the effects of temperature on common forms of food poisoning, such as salmonellosis. These studies show an approximately linear increase in reported cases for each degree increase in weekly temperature. Moreover, there is evidence that temperature variability affects the incidence of diarrhoeal disease. A number of studies found that rising temperatures were strongly associated with increased episodes of diarrhoeal disease in adults and children. These findings have been corroborated by analyses based on monthly temperature observations. Several studies report a strong correlation between monthly temperature and diarrhoeal episodes on the Pacific Islands, Australia, and Israel. Extreme rainfall events can increase the risk of outbreaks of water-borne diseases particularly where traditional water management systems are insufficient to handle the new extremes. Likewise, the impacts of flooding will be felt most strongly in environmentally degraded areas, and where basic public infrastructure, including sanitation and hygiene, is lacking. This will raise the number of people exposed to water-borne diseases (*e.g.*, cholera) and thus lower their capacity to effectively use food.

IMPACTS OF CLIMATE CHANGE ON ACCESS TO FOOD

Access to food refers to the ability of individuals, communities, and countries to purchase sufficient quantities and qualities of food. Over the last 30 years, falling real prices for food and rising real incomes have led to substantial improvements in access to food in many developing countries. Increased purchasing power has allowed a growing number of people to purchase not only more food but also more nutritious food with more protein, micronutrients, and vitamins. East Asia and to a lesser extent the Near-East/ North African region have particularly benefited from a combination of lower real food prices and robust income growth. From 1970 to 2001, the prevalence of hunger in these regions, as measured by FAO's indicator of undernourishment, has declined from 24% to 10.1% and 44% to 10.2% respectively. In East Asia, it was endogenous income growth that provided the basis for the boost in demand for food, which was largely produced in the region; in the Near-East North African region demand was spurred by exogenous revenues from oil and gas exports, and additional food supply came largely from imports. But in both regions, improvements in access to food have been crucial in reducing hunger and malnutrition.

FAO's longer-term outlook to 2050 suggests that the importance of improved demand side conditions will even become more important over the next 50 years. The regions that will see the strongest reductions in the prevalence of undernourishment are those that are expected to see the highest rates of income growth. South Asia stands to benefit the most. Spurred by high income growth the region is expected to reduce the prevalence of undernourishment from >22% currently to 12% by 2015 and just 4% by 2050. Progress is also expected for sub-Saharan Africa, but improvements will be less pronounced and are expected to set in later. Over the next 15 years, for instance, the prevalence of undernourishment will decline less than in other regions, from H"33% to a still worrisome 21%, as significant constraints (soil nutrients, water, infrastructure, etc.) will limit the ability to further increase food production locally, while continuing low levels of income rule out the option of importing food. In the long run, however, sub-Saharan Africa is expected to see a more substantial decline in hunger; by 2050, <6% of its total population are expected to suffer from chronic hunger. It is important to note that these FAO projections do not take into account the effects of climate change. By coupling agro-ecologic and economic models, others have gauged the impact of climate change on agricultural gross domestic product (GDP) and prices. At the global level, the impacts of climate change are likely to be very small; under a range of SRES and associated climate-change scenarios, the estimates range from a decline of "1.5% to an increase of +2.6% by 2080. At the regional level, the importance of agriculture as a source of income can be much more important. In these regions, the economic output from agriculture itself will be an important contributor to food security. The strongest impact of climate change on the economic output of agriculture is expected for sub-Saharan Africa, which means that the poorest and already most food-insecure region is also expected to suffer the largest contraction of agricultural incomes. For the region, the losses in agricultural GDP, compared with no climate change, range from 2% to 8% for the Hadley Centre Coupled Model, version 3 and 7–9% for the Commonwealth Scientific and Industrial Research Organisation projections.

IMPACTS ON FOOD PRICES

Essentially all SRES development paths describe a world of robust economic growth and rapidly shrinking importance of agriculture in the long run and thus a continuation of a trend that has been underway for decades in many developing regions. SRES scenarios describe a world where income growth will allow the largest part of the world's population to address possible local production shortfalls through imports and, at the same time, find ways to cope with safety and stability issues of food supplies. It is also a world where real incomes rise more rapidly than real food prices, which suggests that the share of income spent on food should decline and that even high food prices are unlikely to create a major dent in the food expenditures of the poor. However, not all parts of the world perform equally well in the various development paths and not all development paths are equally benign for growth. Where income levels are low and shares of food expenditures are high, higher prices for food may still create or exacerbate a possible food security problem.

There are a number of studies that have ventured to measure the likely impacts of climate change on food prices. The basic messages that emerge from these studies are: first, on average, prices for food are expected to rise moderately in line with moderate increases of temperature (until 2050); some studies even foresee a mild decline in real prices until 2050. Second, after 2050 and with further increases in temperatures, prices are expected to increase more substantially. In some studies and for some commodities (rice and sugar) prices are forecast to increase by as much as 80% above their reference levels without climate change. Third, price changes expected from the effects of global warming are, on average, much smaller than price changes from socio-economic development paths. For instance, the SRES A2 scenario would imply a price increase in real cereal prices by H"170%. The (additional) price increase caused by climate change (in the Hadley Centre Coupled Model, version 3, climate change case) would only be 14.4%. Overall, this appears to be the sharpest price increase reported and it is not surprising that this scenario would imply a stubbornly high number of undernourished people until 2080. However, it is also needless to say that a constant absolute number of undernourished people would still imply a sharp decline in the prevalence of hunger, and, given the high population assumptions in the SRES A2 world (>13.6 billion people globally and >11.6 billion in the developing world) this would imply a particularly sharp drop in the prevalence from currently 17% to H"7% by 2080.

IMPACT OF CLIMATE CHANGE IN INDIAN AGRICULTURE

The Food and Agricultural Organization (FAO) has warned that India could lose upto 125 million tones of cereals. The Intergovernmental Panel on Climate Change (IPCC) which was awarded Nobel Prize in the year 2008 has warned that due to global warming, the sea levels in Asia will rise by atleast 40cm by 2100 flooding vast areas on the coastline, including some of the most densely populated cities, whose population will be forced to migrate inland (The Hindu Business Line dated 03-11-2007). The first to be affected by the climate change will be those living in the fringe zones, working outdoors or whose livelihoods depend directly on the physical environment. Viewed from these perspectives, the main sectors facing risks from direct impact could be agriculture, water resources and health.

The IPCC report indicates high probability of crop losses with increase in temperature in the tropical regions. Studies do confirm this trend. Among cereal crops important for food security, wheat is most sensitive to even small increase in temperature. Relatively, rice has greater tolerance to temperature increases. Increasing climate variability could result in considerable season/annual fluctuation in food production. All agricultural commodities are subject to such variability.

SOLUTIONS FOR GLOBAL WARMING

The most important contributor to global warming is the increase in atmospheric CO_2 levels due to human activity. The following are the some of the suggestions to prevent global warming.

CO_2 emission can be cut by reducing the use of fossil fuels by cutting back on car use, investing in energy efficiency, implementing energy conservation measures and utilizing renewable resources such as wind, solar and hydropower. Reducing fossil fuel user will, coincidentally, also reduce emissions of methane, nitrous oxide and ozone.Since CO_2 is consumed by plants and trees (known as carbon-sinks), reversing deforestation and implementing reforestation programmes may reduce levels in the atmosphere. The need to take such measures was first recognized at the RIO climate change convention in 1992, after which over 160 countries pledged to limit emissions of CO_2 and to protect and enhance natural 'sinks' of CO_2. Trees play a unique role in the global carbon cycle. They are the largest land-based mechanism for removing CO_2 from the air. Trees are able to store a large amount of CO_2 in their structure. An acre of forest will absorb about ten times the CO_2 amount absorbed by an acre of crop land. Planting more trees and reducing more trees and reducing timber cut worldwide will help to restore the balance.

Try to cut down demand for electricity when we reduce electricity power use, we save money, breathe cleaner air and help to reduce the global warming problem. Government should also encourage the use of mass transit; provide tax rebates for people who buy low and no-pollution vehicles and subsidies to fossil fuel and the nuclear industries.

Green house gas emissions be managed using an incentive based policy, such as market based approach to capping and reducing such emissions. This type of strategy provides clear incentives for changes in business practices and the development of new technologies. It assures that economic forces are directed to keeping the cost of reducing emissions as low as they can be. Many industrial nations have now adopted policies intended to limit green house gases.

Try to follow the environment-friendly policies like 'Reduce', 'Reuse' and 'Recycle'. Save containers, bags, everything that can be used in the future. Also we use cloth towels and napkins instead of paper ones and use rechargeable batteries instead of disposable ones.

Try to minimize the usage of air conditioners, refrigerators and other electronic goods which cause hazardous effects to the mother earth and can maximize the usage of the solar energy and wind energy for the power generation.

PLANT DIVERSITY IN THE INDIAN GENE CENTRE

India is located between 8°-38°N and 68°-97°.5' E and exhibits extreme variation in altitude-from sea level to heights above vegetational limits in the Himalayas (ca 4500 m), and climate-from monsoon tropical in south to temperate and alpine in the north-western Himalayas and extremely arid to semi-arid in the north-western plains. It is floristically extremely rich with about 33 percent of its botanical wealth (over 15,000 species of higher plants) being endemic. There are about 141 endemic genera distributed over 47 families (Nayar, 1980). Further, of the 4,900 endemic species, larger percentage is localised in the

Himalayas (about 2,532 species) than in other regions, namely, the peninsular tract (1,788 species), and the Andaman and Nicobar Islands (185 species). It is also estimated that floristic richness is maximum in the north-eastern region, which holds about 50 percent of India's total species diversity, *i.e.*, more than 7,000 species, and is considered as the cradle of flowering plants. Of 990 species of orchids, 700 species occur in this region (Nayar, 1989). This, thus, makes the Indian region botanically unique and interesting.

On the basis of distribution of flora, the Indian subcontinent has been divided into 8 to 10 regions. The sub-regions suggested by Chatterjee (1939) are: (1) Western Himalayas, (2) Eastern Himalayas, (3) Central Himalayas, (4) Assam, (5) Gangetic plain, (6) Indus plain, (7) Deccan, and (8) Malabar. Based on physiographic, climatic and cultural features, the ICAR recognised 8 agro-climatic regions (Murthy and Pandey, 1978), while recently, further micro-climatic considerations pin-point to 21 such regions. Also, the Planning Commission has demarcated the country into 15 agro-climatic zones. However, in the present account, broadly 8 regions have been recognised which more or less superimpose over the phyto-geographical regions of Chatterjee (1939). These plant/crop diversity regions, the zones proposed by the Planning Commission.

Enough evidence exists pointing to the range of crops of the Harappans, and the appearance of new crops later in the archaeological record, to date the evolutionary process in Indian crops within reasonable limits. The crops of the Harappan period were chiefly of West Asian origin. They included wheat, barley and peas. Of indigenous Indian origin were rice, tree cotton and probably sesame. Rice first appeared in Gujarat and Bihar, not in the centre of the Harappan culture in the Indus Valley. There is some rather doubtful evidence that African crops were also grown by the Harappans. There is a record of sorghum (*jowar*) from Sind and of *Pennisetum* (*bajra*) from Gujarat. Sesame recorded from Punjab is of uncertain origin, having been ascribed by some to Africa and by some to India. The earliest record of the African cereal, *Eleusine coracana* (*ragi*) is from Mysore, about 1800 B.C. The South-East Asian crops of importance to India are sugarcane and banana, and they are not of the type likely to leave identifiable remains in the archaeological record. Consequently, nothing is known of them before historic times but both appear in the early literary record. Crops of American origin include maize, grain amaranths and potato. The dating of the introduction of maize is uncertain, the characteristics and distribution of some forms being such as to lend support to the view that they reached India in pre-Columbian times. The potato, on the other hand, was certainly a late introduction by either the Portuguese or the British.

The oldest crops have been under the selective pressures of the Indian environment for about 4500 years, though most of the crops except those of American origin have been exposed for at least 2000 years. The post-Columbian introductions from America, on the other hand, have been in India for less than 500 years.It may also be emphasised here that crops of the Indian subcontinent have influenced the agricultural development of ancient Egyptian, Assyrian, Sumerian and the Hittite Civilization (Nile/Tigris/Euphrates river basins) through their early spread to these regions of the Old World. Also during the early Christian era, several Indian crops/plants were taken to South-East Asian countries by the Buddhists. Much early exchange of plant material had also existed with Africa, while many crops, *viz. Citrus*, cotton, jute, rice and sugarcane were distributed especially to the Mediterranean region by the Arabs in the 8-10th centuries A.D.

Similarly, several New World domesticates were introduced into India and reciprocally Indian cultigens were subsequently introduced into the New World (Mehra and Arora, 1982). Further, prevalence of great ethnic diversity and primitive agriculture (Arora, 1987) also points out to the rich heritage of the Indian subcontinent.

INDIAN SUBCONTINENT AS GENE CENTRE

India is one of the twelve centres/regions of diversity of crop plants in the world. As is evident from the above brief account, antiquity of agriculture as also the ethnic diversity in the sub-continent have played a major role in the diversification of crop resources in this region. Thus, rich genetic diversity occurs in several crop plants and their wild progenitors. As it is, the diverse agro/eco-climate of the 'Hindustani Centre', apart from possessing about 166 species of agri-horticultural crop plants, also has a rich diversity in wild relatives of crop plants numbering about 320 species. In addition, as already reflected above, the Indian agriculture has been enriched by a continuous stream of introductions of new crops and their cultivars by man since the ancient times.

The Mughals, Spaniards and the Portuguese brought several new crops to this country. Prominent amongst these were pear, grape, nut fruits, date palm, garlic, opium, maize, potato, sweet potato, tomato, chillies, French bean, peas, etc. The Arabs possibly brought clove, coriander, cumin, ammi or lavage (*ajwain*) and fennel. This process continued during the British period which saw the introduction of coffee, cocoa, cashew nut, strawberry, litchi, *Cinchona*, etc., besides collection of tea in Assam and adjoining region. The current matrix of diversity consists of the genepool of indigenous crop plants, their wild and/or weedy relatives and the well adapted introductions from practically all over the globe. The introduced types, depending on the time of introduction, extent of material introduced and the areas of introduction, exhibit enormous build-up of diversity within the Indian subcontinent.

Some good examples are: *cereals*-wheat, barley, oats, maize; *pulses*-gram, French bean (*rajmah*) and peas; *vegetables*-potato, onion, cauliflower, cabbage, carrot and tomato; *fruits*-apple, pear, grapes, cherry, peach and apricot; *oilseeds*-

soybean, sunflower and groundnut; *fibre crop*-cotton; to a very limited extent also in *medicinal plants*-mint, liquorice, *Digitalis* (foxglove), *Cinchona* (quinine), *Hyoscyamus* (henbane) and others such as *Humulus lupulus* (hops).

GENETIC DIVERSITY AND ITS DISTRIBUTION

Our knowledge on the origin of cultivated plants and regions of diversity has considerably increased since the time of de Candolle (1886) and Vavilov (1926). This has been amply elaborated. By and large, these geographic areas vary from eight to twelve in number, lie between 20° and 45° latitude north and south, and are superimposed over the world's major vegetation formations/biomes/floristic zones (Good, 1953). The Indian gene centre is one such important region of diversity of crop plants. It has strong linkages and contiguity with the Indo-Chinese-Indonesian, Chinese-Japanese and the Central-Asian regions. Further, as pointed out above, the influx of germplasm in distant past from the Mediterranean, African and tropical American regions, has built up enormous locally adapted and selected variability.

It has already been stressed that the Indian subcontinent possesses rich floristic wealth. Of this 3,000-5,000 species are of economic value which include about 1,000 wild food plants. The diversity in plant genetic resources numbers around 166 species and the wild genepools of direct or indirect utility, around 320 species, distributed in eight phyto-geographical/agro-ecological zones. By and large, India is:

1. a primary centre of diversity for crops, such as rice, black gram, moth bean, pigeonpea, cucurbits like smooth gourd, ridge gourd and pointed gourd, tree cotton, *capsularis* jute, jack fruit, banana, mango, *Syzygium cumini*/jamun, large cardamom, black pepper and several minor millets and medicinal plants like *Rauvolfia serpentina* and *Saussurea lappa.*
2. a secondary centre of diversity for African crops, such as finger millet, sorghum, cowpea, cluster bean, okra, sesame, niger and safflower; tropical American types like maize, tomato, pumpkin/*Cucurbita* spp., chayote or chou chou, chillies and *Amaranthus;* and with,
3. regional (Asiatic) diversity, for crops like maize, barley, amaranth, buckwheat, prosomillet, foxtail millet, mung bean/green gram, chickpea, cucumber, bitter gourd, bottle gourd, snake gourd and Brassicae.

Geographical contiguity with the Far-East and/or the Indo-Malayan (South/South-East Asian region) belt is largely responsible for more regional diversity in mung bean, rice bean, sword bean, tomato, citrus, small cardamom, *Saccharum*, ginger, turmeric, tuber crops, particularly taros and yams, and bamboos.

EXTENT OF PREVALENT DIVERSITY

Different agro-ecological/phyto-geographical regions of India, thus hold rich diversity in both the cultivated and the wild crop genepools. Based on the areas and crops, the following brief analysis on the distribution and extent of crop-plant diversity is given:

CROP PLANTS

The following regions exhibit richness in crop plant diversity:

Western Himalayas including cold arid tract: In Kinnaur, Lahaul and Spiti, Pangi, Ladakh and adjoining areas, rich genetic diversity occurs in wheat, barley (particularly hull-less types), buckwheat, prosomillet, amaranth, chenopods, field peas, lentil and several other crops, possessing adaptability to cold and tolerance to drought conditions. In barley, bluish/black grain types occur at high altitudes (above 3300 m), exhibiting akinness to local Tibetan types. In wheat, both awned and awnless, tall types occur and collections from Lahaul and Spiti have exhibited resistance/tolerance to rust. Further, rich diversity in rice, maize, French bean, peas, buckwheat, amaranth and other crops-radish, leafy Brassicae, pome, stone and nut fruits, forages and medicinal plants, occurs in the sub-humid hill regions of Jammu and Kashmir, Himachal Pradesh and Uttar Pradesh.

North-eastern region and the Eastern Himalayas: The tribal dominated belts of Mizoram, Manipur, Meghalaya, Tripura, Sikkim, North Bengal and parts of Nagaland and Arunachal Pradesh, are rich in local variability of cereals-rice, maize (including the primitive pop-corn), barley, wheat; pseudocereals-buckwheat, *Chenopodium*, amaranth and soft-shelled forms in *Coix;* millets-finger millet, foxtail millet; legumes-rice bean, winged bean, adzuki bean, black gram, sem/*Dolichos*, soybean, sword bean, peas; oilseeds-*Brassica* spp., *Perilla*, sesame, niger; vegetables-cucurbits (*Cyclanthera*, *Cucurbita*, *Momordica*, *Cucumis*, *Luffa*, *Lagenaria*, *Benincasa*); fibres-*Corchorus* spp., tree cotton, kenaf and mesta; taro/yam and several other tuberous/rhizomatous/bulbous types, and in *Citrus*, *Musa* and bamboos.

Eastern peninsular region: The tribal dominated tract of Orissa and the Chotanagpur plateau may serve as good examples of areas that hold rich crop diversity *viz.* rice, sorghum, finger millet, fox-tail millet, prosomillet and other minor millets, *Dolichos*, rice bean, chickpea, pigeonpea, horsegram/kulthi, sesame, niger, Brassicae, linseed, castor, several cucurbits, brinjal, chillies, mango and wild types such as in rice (*Oryza nivara*).

Western arid/semi-arid region: In Rajasthan and Gujarat (including Saurashtra), rich diversity occurs in sorghum, pearl millet, moth bean, cluster bean/*guar*, cowpea, black gram, green gram, Brassicae, sesame, cucurbits, forage legumes and grasses.

Central tribal region: Madhya Pradesh and adjoining tract of Maharashtra possess rich diversity in wheat, rice, sorghum, minor millets, grain legumes-particularly cowpea, chickpea, pigeonpea, black gram and green gram; oilseeds-niger, sesame, Brassicae; and cucurbits.

Western peninsular region: The Western Ghats tract has promising diversity in tuber crops-*Colocasia, Dioscorea* and *Coleus;* rhizomatous types like *Curcuma* species and ginger, and other crops like rice, finger millet, several grain legumes, and much variation in cucurbits, forage legumes and grasses; several fruit trees such as jack fruit and spices like cardamom and black pepper.

CROP DIVERSIFICATION IN THE INDIAN PERSPECTIVE

With the advent of modern agricultural technology, especially during the period of the Green Revolution in the late sixties and early seventies, there is a continuous surge for diversified agriculture in terms of crops, primarily on economic considerations. The crop pattern changes, however, are the outcome of the interactive effect of many factors which can be broadly categorized into the following five groups:

- Resource related factors covering irrigation, rainfall and soil fertility.
- Technology related factors covering not only seed, fertilizer, and water technologies but also those related to marketing, storage and processing.
- Household related factors covering food and fodder self-sufficiency requirement as well as investment capacity.
- Price related factors covering output and input prices as well as trade policies and other economic policies that affect these prices either directly or indirectly.
- Institutional and infrastructure related factors covering farm size and tenancy arrangements, research, extension and marketing systems and government regulatory policies.

Obviously, these factors are not watertight but inter-related. For instance, the adoption of crop technologies is influenced not only by resource related factors but also by institutional and infrastructure factors. Similarly, government policies-both supportive and regulatory in nature-affect both the input and output prices. Likewise, special government programmes also affect area allocation and crop composition. More importantly, both the economic liberalization policies as well as the globalization process are also exerting strong pressures on the area allocation decision of farmers, essentially through their impact on the relative prices of inputs and outputs. Although the factors that influence the area allocation decision of farmers are all important, they obviously differ in terms of the relative importance both across farm groups and resource regions. While factors such as food and fodder self-sufficiency, farm size, and investment constraints are important in influencing the area allocation pattern among smaller farms, larger farmers with an ability to circumvent resources constraints usually go more by economic considerations based on relative crop prices than by other non-economic considerations. Similarly, economic factors play a relatively stronger role in influencing the crop pattern in areas with a better irrigation and infrastructure potential. In such areas, commercialization and market networks co-evolve to make the farmers more dynamic and highly responsive to economic impulses.

What is most notable is the change in the relative importance of these factors over time. From a very generalized perspective, Indian agriculture is increasingly getting influenced more and more by economic factors. This need not be surprising because irrigation expansion, infrastructure development, penetration of rural markets, development and spread of short duration and drought resistant crop technologies have all contributed to minimizing the role of non-economic factors in crop choice of even small farmers. What is more, the reform initiatives undertaken in the context of the ongoing agricultural liberalization and globalization policies are also going to further strengthen the role of price related economic incentives in determining crop composition both at the micro and macro levels. Obviously, such a changing economic environment will also ensure that government price and trade policies will become still more powerful instruments for directing area allocation decisions of farmers, aligning thereby the crop pattern changes in line with the changing demand-supply conditions. In a condition where agricultural growth results more from productivity improvement than from area expansion, the increasing role that price related economic incentives play in crop choice can also pave the way for the next stage of agricultural evolution where growth originates more and more from value-added production.

CONSEQUENCES OF CROP PATTERN CHANGES

Turning now to the socio-economic and environmental consequences of crop pattern changes, the Green Revolution technologies have fomented, among other things, an increasing tendency towards crop specialization and commercialization of agriculture. While these developments have positive effects on land/labour productivity and net farm income, they have also endangered a number of undesirable side effects like reduced farm employment and crop imbalances. Although the expansion of commercialized agriculture has fomented new sets of rural non-farm activities and strengthened the rural-urban growth linkages, it has also weakened the traditional inter-sectoral linkages between the crop and livestock sectors. Besides, crop pattern changes also lead to serious environmental consequences that take such forms as groundwater depletion, soil fertility loss and waterlogging and salinity-all of which can reduce the productive capacity and growth potential of agriculture over the long-term. A classical example is the rice-wheat system in Northwestern India replacing traditional crops like pulses, oilseeds and cotton.

CROP PATTERN CHANGES: A TEMPORAL ANALYSIS

The temporal analysis of the changes in crop pattern is observed both at the national and state level. For the purpose of the present study, crop pattern changes at the all India level are evaluated by considering the area share of crops and crop groups at four time points capturing, respectively, the triennium ending (The) averages of areas in 1966/67, 1976/77, 1986/87 and 1996/97. These time points have been selected so as to capture the major events and stages in the evolution of Indian agriculture which are of direct relevance to the purpose of this study.

The pre-Green Revolution situation is represented by the 1966/67 period. The post-Green Revolution situation is represented, on the other hand, by three sub-periods. These periods have been chosen deliberately to highlight the temporal differences in the factors contributing to agricultural growth. In the first of these three post-Green Revolution periods, *i.e.*, 1966/67 to 1976/77, the impact of the Green Revolution has mainly been on wheat, especially in the irrigation-wise better endowed Northwestern parts of India. But in the second period, *i.e.*, 1976/77 to 1986/87, the new technologies have expanded to other regions and also covered crops such as rice. During this period, the Southern and Eastern regions showed remarkable improvement in agricultural growth and productivity. More or less during the same period, the benefits of the new crop and irrigation technologies that were confined earlier to large and medium sized farms have also started percolating to the small farms. The third period, *i.e.* 1986/87 to 1996/97, is characterized by a number of policy changes including the launching of the Technology Mission on Oilseeds (TMO) as well as price support and stabilization policies for oilseed crops. During this period there is also a marked shift in the pattern of agricultural investment in favour of the private sector. This period not only covers such important policy changes as unleashed by agricultural liberalization policies but also represents the consolidation phase of rural infrastructure. Since these four periods can capture well the changes in the set of factors that affect crop pattern, they can provide a better analytical framework for evaluating both the nature and technical context of crop pattern shifts observed not only at the national level but also at the state level.

ANALYSIS AT THE ALL INDIA LEVEL

The analysis of crop pattern changes to be attempted at the macro level (national) will focus on three main aspects. These aspects are: a) the nature and direction of area shifts across crops and crop groups observed through time, b) the implications of these shifts for crop diversification and balance in the inter-crop allocation of existing and additional areas brought under cultivation, and c) the output and productivity impact of crop pattern changes.

CROP DIVERSIFICATION IN INDIA

India is a country of about one billion people. More than 70 percent of India's population lives in rural areas where the main occupation is agriculture. Indian agriculture is characterized by small farm holdings. The average farm size is only 1.57 hectares. Around 93 percent of farmers have land holdings smaller than 4 ha and they cultivate nearly 55 percent of the arable land. On the other hand, only 1.6 of the farmers have operational land holdings above 10 ha and they utilize 17.4 percent of the total cultivated land. Due to diverse agro-climatic conditions in the country, a large number of agricultural items are produced. Broadly, these can be classified into two groups-foodgrains crops and commercial crops. Due to the challenge of feeding our vast population and the experience of food shortages in the pre-independence era, 'self reliance' in foodgrains has been the cornerstone of our policies in the last 50 years. Around 66 percent of the total cultivated area is under foodgrain crops (cereals and pulses). Concurrently, commercial agriculture developed for whatever reasons in the pre-independent phase also kept flourishing during the post independent period. Commercial agriculture not only catered to the domestic market but has also been one of the major earners of foreign exchange for the country.

Crop diversification is intended to give a wider choice in the production of a variety of crops in a given area so as to expand production related activities on various crops and also to lessen risk. Crop diversification in India is generally viewed as a shift from traditionally grown less remunerative crops to more remunerative crops.

The crop shift (diversification) also takes place due to governmental policies and thrust on some crops over a given time, for example creation of the Technology Mission on Oilseeds (TMO) to give thrust on oilseeds production as a national need for the country's requirement for less dependency on imports. Market infrastructure development and certain other price related supports also induce crop shift. Often low volume high-value crops like spices also aid in crop diversification. Higher profitability and also the resilience/stability in production also induce crop diversification, for example sugar cane replacing rice and wheat. Crop diversification and also the growing of large number of crops are practiced in rainfed lands to reduce the risk factor of crop failures due to drought or less rains. Crop substitution and shift are also taking place in the areas with distinct soil problems. For example, the growing of rice in high water table areas replacing oilseeds, pulses and cotton; promotion of soybean in place of sorghum in vertisols (medium and deep black soils) etc.

TEMPORAL CHANGES IN THE AREA SHARE OF CROPS

The temporal behaviour of crop pattern changes at the all India leve, respectively, the area share of main crop groups and major crops for the four periods. Though obvious, it needs to be stated that the changing area share of crops is due as much to shift in area under other competing or alternative crops as to the relative area allocation of fresh areas brought under cultivation. In any case, the changing area share of crops does capture the ongoing changes in the comparative advantage calculus of farmers. The changes in the comparative advantage of crops reflect, in reality, the ongoing changes in relative prices of inputs and outputs, production conditions (including irrigation expansion), development and spread of new crop and farm technologies, extension and input support policies and trade policies and domestic regulations. As such, the changing area share of crop pattern, though looking deceptively simple, becomes a useful tool for understanding the direction in which crop pattern changes are influenced by the variations in the comparative advantage of crops and crop groups.

Although cereals gained a marginal increase in area share in the first decade of the Green Revolution, their area and share declined gradually thereafter. Between 1966/67 and 1996/97, 3.35 percent of the gross cultivated area (GCA)-representing approximately about 5.7 million hectares (m/ha)-has shifted from cereal crops to non-cereal crops. Since the area share of pulses taken as a group also declined by 1.57 percent during the same period, the area share of foodgrains as a group declined by 4.92 percent during 1966-97. In area terms, the shift from foodgrains to non-foodgrains involves an approximate area of about 8.36 m/ha. While cereals and pulses have lost area, the major gainers of this area shift are the non-foodgrain crops especially oilseeds. The area share of oilseeds as a group that has gone up by 4.08 percent accounts for about 83 percent of the 8.36 m/ha involved in the area shift between 1966/67 and 1996/97. As we consider the share of individual crops within cereals, although the share of cereals as a group has declined, the area share of rice has increased continuously over all the four periods. Wheat, although having a declining area share until 1986/87, also gained in its share when the entire period is considered. Thus, the area loss of cereals can be attributed entirely to the declining area share of coarse cereals, especially sorghum, pearl millet, barely and small millets. It can be noted that even within coarse cereals, the area share of maize shows a marginal improvement over the years. Within oilseeds, the crops showing steady improvement in their area share are: rapeseed and mustard, soybean and sunflower.

Among these three oilseeds gaining in area share, rapeseed and mustard are substantially grown as intercrops with wheat. On the other hand, the area shares of other oilseeds including groundnut (that has a dominant area share within oilseeds) but excluding coconut, which is more a plantation crop than field crop, have either fluctuated or declined. The area share of groundnut, though improved during the last period, has declined as compared to its share in the pre-Green Revolution period. But, the declining area share of crops-especially those with only a marginal change in their area share-need not necessarily imply a decline in the actual area under these crops. Since the Gross Cropped Area (GCA) is constantly increasing over time, partly through an expansion of net sown areas as in the initial stages of the Green Revolution and partly through increasing intensity of cropping mainly by irrigation expansion, the declining area share can coincide with an increase in absolute increase in the area under crops. Although the increase in the area share of other commercial crops is not as dramatic as that of oilseeds, it is still notable because of its implications for the direction of Indian agriculture.

But, among these other commercial crops that cover fibres, spices, fruits and vegetables, and other field crops such as tobacco and sugar cane and plantation crops, only spices, fruits and vegetables show a steady improvement in their area shares, whereas others show mostly a declining trend. This is particularly true for fibres and other field crops that have over four fifths of the total area under the broad group of other commercial crops. However, sugar cane, included in the category of other field crops, shows an increase in its area share. This is also true for cotton included in the fibre category. While all spice crops show a gradual increase in their area share, only three of the six crops included in the fruits and vegetables category show a gain in their area share over the years. These crops are banana, potato and onion.

SUCCESS STORY IN CROP DIVERSIFICATION

It is clear that most of the area shifts that occurred during the three decades between 1966/67 and 1996/97 is from coarse cereals to oilseeds. Three notable aspects of this area shift towards oilseeds can be noted. First, most of these area shifts have occurred particularly during the decade ending 1996/97. A protective trade environment, favourable price policy and the connecting of the Technology Mission on Oilseeds (TMO) during this period have all enhanced the comparative advantage of oilseeds.

The area share of oilseeds that increased just by a percentage each during the two decades before 1986/87 has risen suddenly by 43 percent during the decade ending 1996/97. Second, the area shift during this period comes mainly from barley and pulses other than pigeon pea. Since most of those crops losing their area share are usually grown under rainfed conditions where oilseeds can also be grown, the area shift can be said to involve mostly rainfed areas, although comparative advantage and crop rotation considerations often favour oilseeds even in groundwater irrigated areas.

There is also a significant area shift within oilseed crops. For instance, while the area shares of rapeseed and mustard, sunflower and soybean are increasing steadily, those of sesamum, linseed and nigerseed are declining gradually. Thus, the area shift has favoured only a sub-sector within the oilseed sector partly because of constant changes in the comparative advantage of different oilseeds and partly because of the impact of changing consumers' preferences on the relative demand of oilseeds.

CROP DIVERSIFICATION AND COMPOSITION

As noted already, area shifts and crop pattern changes can lead either to crop specialization or to crop diversification. The area share of foodgrains increased during 1967-76 due partly to their yield advantages created by irrigation expansion and Green Revolution technologies and partly to government policies pursued to encourage food production and eliminate food imports. As a result, there was a tendency towards cereal centered specialization. But, later when increased productivity of foodgrains, especially cereals, made it possible to allocate more area to other crops such as oilseeds with a severe supply shortage, the specialization tendency witnessed earlier has given room for overall crop diversification. Even within such an overall crop diversification, it is also possible to see specialization tendencies within each crop group. For instance, within cereals, the declining share of coarse cereals and increasing share of wheat and rice indicates a rice and wheat centered specialization.

Such a specialization has become possible because within the achievement of self-sufficiency, mainly through an expanding production of wheat and rice, the food implication of coarse cereals has declined almost coinciding with their declining demand prompted by an increasingly incurring income level. On the other hand, within oilseeds, even though groundnut still has a dominant area share, the growth of area under rapeseed and mustard, sunflower and soybean indicates an ongoing structural change leading to diversification within the oilseed sector. The issue of crop diversification, although considered in area terms, assumes significance in view of its effects on the supply-demand balance of main crops and crop groups. The rice and wheat-centered specialization, for instance, indicates an increase in the supply of wheat and rice but a reduced supply of coarse cereals. Since the demand for coarse cereals is declining and that for wheat and rice is increasing due to changes in the income pattern, the changes in their supply are actually necessary to achieve the required demand-supply balance. A similar line of argument can also be extended to other crops where in addition to domestic demand, international demand and supply also assume significance.

Clearly, during the Green Revolution period, the foodgrains category has evinced an increasing tendency towards specialization with an unbalanced area composition, whereas the non-foodgrains category has shown an increasing tendency to diversify with an improved balance in their area composition. But, during 1976-97 the non-foodgrains category has also caught up with the specialization tendency and area imbalance that characterized foodgrains throughout the entire period. The reason for the specialization tendency within foodgrains is rather obvious in view of the declining area share of coarse cereals and the increasing area share of rice and wheat. The changing area share of crops can also explain the specialization tendencies observed among the non-foodgrains. As we have seen in the context, the area shares of oilseeds and other commercial crops have increased especially during the post Green Revolution period characterized by a prices and market-responsive agriculture inevitable in the aftermath of achieving food self-sufficiency. It is important to note, however, that oilseeds are more diversified with an enhanced balance in their area composition. Within foodgrains, on the other hand, while cereals tend towards specialization with an increasing imbalance in their area composition, pulses evince a tendency to diversify with a declining imbalance in their area composition. The latter result is mainly due to the fact that the area shares of the dominant pulses such as gram have declined to give room for other pulses such as pigeon pea. But it should be noted that diversification and area balance achieved by pulses have occurred only with a reduced area share. Among the non-foodgrain crops, spices are notable not only for their increasing tendency towards diversification but also for a declining imbalance in their area composition. In contrast, fibres, fruits and vegetables, other field crops and plantation crops have all evinced a specialization tendency especially after 1986/87.

7

Vegetable Gardening in Crop Plants

VEGETABLE GARDENS

Successful vegetable gardens are not accidental. They are the results of planning, constant care, and the will to make things grow. Among the many things a vegetable garden may offer towards a satisfying experience are fresh air, exercise, sunshine, knowledge, supplemental income, mental therapy, and fresh food, rich in vitamins and minerals, harvested at the best stage of maturity.

Organic gardening differs from "conventional" gardening mainly in the areas of fertilization and pest control. The organic gardener prefers to use natural and organic materials and methods, and avoids using practices and synthetic chemicals that may be detrimental to his health or environment. The information in this publication should be beneficial to all gardeners regardless of methods of culture used; however, it is primarily intended to aid the organic gardener to employ workable methods acceptable to him and compatible with the philosophy of "organic gardening."

SOIL PREPARATION

Since organic fertilizer and soil conditioning materials are slow working in general, they should be mixed into the soil at least three weeks ahead of planting and the soil thoroughly prepared for the seed or transplants. Clumps of unrotted organic materials not only interfere with the seeding operation, but may result in nutrient deficiency and possible soil-borne diseases problems such as "damping-off" of young seedlings.

ORGANIC MATTER

A major basis for organic gardening is the use of abundant quantities of organic material applied to the soil. Usually, it is in the form of animal manures, plant manures, cover crops, compost, or mixed organic fertilizer.

Benefits of Adding Organic Matter

Improves tilth, condition, and structure of soil.

Improves ability of soil to hold water.

Improves ability of soil to hold nutrients.

Improves "buffering" capacity of soil; that is, keeps soil from "over-reacting".

Supports the soil's microbiological activity (or the life of the soil).

Contributes nutrients, both minor and major.

Releases nutrients slowly.

Acids arising from the decomposition of the organic matter help to convert insoluble natural additives such as ground rock into plant-usable forms.

Helps vegetables survive stress, as from nematodes.

Helps dispose of organic waste products.

What Happens to Organic Matter Applied to the Soil or Compost Pile

Under suitable conditions, the organic matter is decomposed by micro-organisms such as fungi, algae, bacteria, molds, and earthworms. In the process, insoluble and unavailable (to plants) nutrients, such as nitrogen, are gradually changed into simple usable products. For example, *nitrogen* is converted from the unusable organic forms to a usable inorganic form through the process called nitrification. Thus, nitrification is the breakdown of protein (organic nitrogen) into ammonia and then nitrate. Some of the organic matter becomes part of the soil humus.

PROPER CONDITIONS FOR NITRIFICATION

First, materials containing nitrogen must be present. There is a great variation in the amount of nitrogen the different organic materials contain. Then certain soil or compost conditions are necessary:

Proper soil acidity (pH) —should be about 7.0; in acid situation below 5.5 it ceases.

Proper temperature of soil — above 50°F.

Good aeration — (does not occur with wet, soggy soil or compost).

Adequate lime for use by micro-organisms and to keep the soil from being acid.

ANIMAL MANURES

Where animal manures are available, they are probably the best source of fertilizer and organic matter for the organic gardener. Use manure which has been aged for at least 30 days, or composted. Manures vary greatly in their content of fertilizing nutrients.

The composition varies according to type, age, and condition of animal; the kind of feed used; the age and degree of rotting of the manure; the moisture content of the manure; and the kind and amount of litter or bedding mixed in the manure. shows average minimal amounts of nutrients to be expected. They may be as high as 4.5% N; 2% P; and 2% K in some cases. Animal manures also provide most of the micro-nutrients needed. Some manure products are composted, rehydrated or mixed with plant litter to enhance their fertility.

Before Planting

Cow, horse, hog — A minimum of 25 pounds per 100 square feet (about 5 tons per acre) of garden soil. For best results, supplement each 25 pounds of manure with 2 to 3 pounds of ground rock phosphate or raw bone meal. Use up to 1 pound per square foot.

Poultry, sheep — Apply at least 12 pounds per 100 square feet (about 3 tons per acre minimum) - or as much as 3 - 4 inches (20 tons per acre).

Other animals—Rabbit, goat, and exotic grasseaters — 10-12 pounds/100 sq ft. minimum, up to 100 lbs. per sq. ft.

After Planting (As a sidedressing) if needed

Cow, horse, hog — Sidedress with up to 5 pounds per 100 square feet of row.

Poultry, sheep — Use up to 3 pounds per 100 square feet of row.

All Types — Broadcast evenly over plot and spade, roto-till or otherwise work into topsoil. Apply three or more weeks before planting. A small amount may be mixed well in the planting hole; however, plant injury may occur with 1 lb. or more placed in the hole.

All Types — Scatter a band of manure down each side of the row. Place each band at the edge of the root zone and work lightly into the soil surface. For individual plants, open a furrow encircling the plant and fill with manure, then cover *mulch* is present, rake it back at the edge of the root zone in order to apply the band of manure, then re-cover with the mulch.

COMPOSTS

Acceptable manure-like organic fertilizer (artificial manure) may be obtained through the process of *composting.* Simply put, *compost* is made by alternating layers of organic materials, such as leaves and kitchen table refuse, with manure, topsoil, lime, organic fertilizer, water, and air, in such a manner that it decomposes, combines, and yields artificial manure.

How is the Compost Pile Made

The *compost* pile is made of convenient size, usually not less that 10 feet square (100 square feet) and 3 to 5 feet high. The top should be left flat or with a slight depression in the centre to catch rain or added water. Too much water eliminates air and slows the decay process. One way suggested in building the pile is to make a layer of leaves, straw, grass clippings, and other organic materials 1 foot deep, wet down and pack. Spread a layer of manure 4 to 6 inches deep over this layer of wet material. Then spread up to 5 pounds of ground rock phosphate or 1 quart of raw bone meal per 100 square feet, and 1 pound of ground limestone. Instead of the rock phosphate and bone meal, you could use 5 pounds of a complete organic fertilizer such as Fertrell per 100 square feet. Also, a layer of topsoil is sometimes used. Then continue to repeat the process until the pile has reached 3 to 5 feet high.

Compost will begin to heat after 2 or 3 days. Keep it moist, but not too wet, and do not disturb for awhile. After 3 to 4 weeks, fork it over, mixing the parts to obtain uniformity. Fatty animal wastes tend to create bad odors, draw flies and ants; so, try to avoid their use if this will be a problem. Compost for the garden should be ready from 2 months to 1 year, depending on the time of year, type of materials added, and skill of the composter. When the compost is broken down into a homogenous mixture, and no undecomposed leaves or other material may be seen, it is ready for use.

Organic Materials Used in the Compost Pile

Most anything organic, but most popular materials are natural materials such as straw, leaves, pine straw, grass clippings, shrub clippings, garbage, fish scraps, water hyacinths, etc. A list of materials and what each might contribute to the compost is given in. In addition, certain of the materials listed as organic fertilizers (for example, tobacco stems) could also be added.

Irrigation

In irrigating the garden, it is advisable to thoroughly wet the soil once a week unless sufficient rain falls. Thus, the soil will be moistened throughout the root zone. Light sprinklings every day merely tend to wet the surface and encourage shallow root growth. Drip or trickle irrigation is encouraged as a method for conserving water. Use of organic materials as soil conditioners and fertilizers tends to improve the ability of the soil to retain moisture. Also, a good garden mulch will conserve soil moisture. Please observe all local watering regulations.

Mulching

A mulch is any material, usually organic, which is placed on the soil *surface* around the plants. Organic materials most commonly used for mulching are leaves, grass clippings, pine straw, sawdust, and wood shavings. Synthetic materials, mostly plastic sheeting, have been used quite often in recent years. Among the benefits of a mulch are (a) conserves soil moisture, (b) conserves nutrients, (c) reduces soil erosion, (d) reduces crop loss due to nematodes, (e) reduces weed growth, (f) provides barrier between fruit and soil, thus reducing soil rot on fruit, and (g) moderates the soil temperature.

Apply mulch before or after seeding or transplanting. Roll back the mulch with a rake in order to wet the soil beneath when irrigating, for best results. At the end of the garden season, the mulch (except plastic) may be removed and composted, or cut into the garden soil. Most mulch is woody and should have manure or other rich organic fertilizer applied with it when cutting into the soil.

Weed Control

The primary purpose of cultivation is to control weeds. Weeds are easy to control when they are small. Shallow cultivation and hoeing are advised in order to reduce damage to the root system. A garden mulch, such as pine straw, leaves, or other material, will help to keep weeds from growing if the mulch is thick enough to exclude light.

Use of Compost in the Garden

Since compost is artificial manure, it should be used much as you would manure. Broadcast it over the entire garden three weeks or more before planting. Or if you have only a small quantity of compost, it may be mixed into the soil along each planting furrow or at each hill site. As a minimum, apply it at the rate of about 25 pounds per 100 square feet, or ¼ pound per square foot. Larger amounts are even more beneficial, up to 200 lb/100 sq ft. (2 lb/sq ft.).

Caution: If your compost is made from mostly woody materials, it may temporarily deplete the nitrogen from the soil and plants. Be sure to mix manure with it when applying.

Natural and Organic Fertilizers

Natural and organic materials which yield plant nutrients upon decomposition are often available for purchase either separately or in combination. These materials may be applied to the garden separately or combined, used in the compost pile, or mixed with manure. Many of the more commonly available materials are listed in along with the nutrients which they produce. These include both the organic materials derived from plants and animals, plus the natural deposits of rocks and minerals.

Natural Deposits (Rocks, Sands, Shells, etc.)

Such naturally occurring materials are usually not easily obtained in today's modern agriculture; however, where available they represent sources of mainly potash, phosphorus, and lime (calcium and magnesium) for organic gardeners.

Phosphorus — Rock phosphates are natural deposits of phosphate in combination with calcium. The material as dug from the earth is very hard and yields its phosphorus very slowly. When finely ground and with impurities removed, the powdery material is only slightly soluble in water, but may be beneficial to plants in subsequent seasons following application. The reaction of phosphate rock with acids from decaying organic matter in the garden or compost tend to make the phosphorus available to garden plants. Collodial phosphate is also available and widely used. Apply both phosphates at the rate of 2-5 pounds per 100 square feet of garden soil. Or, when applying manure or compost, mix at the rate of 2½ pounds phosphate per 25 pounds manure or compost. Broadcast the material over the soil surface and work into the topsoil at least three weeks before planting. Manure or other organic fertilizer should be added at this time. Since the materials are so slowly decomposed, sidedressings are seldom beneficial.

Potash — Potassium is widely distributed in nature, occurring in rocks, soils, tissues of plants and animals, and water of seas and lakes. In gardening practice, materials such as wood ashes, tobacco stems,wool suint, seaweed, potash salts, greensand, and ground rock potash are used alone, in combinations with other materials yielding other nutrients, mixed with manure, or in compost piles. Since the potash bearing materials vary so much in composition and rate of decomposition, specific application rates must be determined for each material and its combinations. In general, ground rock potash at 5 pounds per 100 square feet may be broadcast over the soil surface three weeks prior to planting and spaded in. Langbeinite (Sul-Po-Mag) is used at 1 lb/100 sq ft.

Micro-nutrients — An advantage for using organic materials as fertilizers is that they contain many of the elements also needed by the plants in addition to N, P, and K (for example, manganese in manure). Besides the general amounts of micronutrients found in most organic materials, certain ones are concentrated into such naturally occurring materials as gypsum (calcium and sulfur), marl (calcium), dolomite (Calcium and magnesium), limestone (calcium), basic slag (iron, calcium, manganese and magnesium), and finely ground borosilicates.

Lime — Reducing the acidity of the soil is the primary purpose for using lime in the garden. However, liming materials also provide nutrients for plant use. Calcium and magnesium are the two elements most commonly provided by lime. Gypsum is used where more calcium is needed without raising the pH. Natural deposits of lime which are an organic gardener might use are limestone, dolomite, shell, and marl. All these forms must be finely ground to provide maximum benefit to the soil and plants. Dolomite is preferred due to its ocntent of both calcium and magnesium. Lime to sweeten the soil should be applied only when the needs have been established by a reliable soil test. Under most Florida soil conditions, applications of 2 to 5 pounds of finely ground dolomitic limestone per 100 square feet usually will be sufficient except on very acid soils. Apply lime well in advance of the planting date, preferably 2 to 3 months before the garden is planted. Mix well with the soil and keep moist for best reaction. Application closer to planting time is permissible, but its benefits are delayed.

Insect and Disease Control

During periods when infestations of various garden pests are high, control by natural means becomes very difficult. However, the following practices will help to reduce losses without use of chemical pesticides.

Plant Resistant Varieties

Plant seed from disease-free plants.

Select pest-free transplants.

For cutworms, place a cardboard of tinfoil collar around plant stems at ground level.

Spade garden early so vegetation has time to rot before planting.

Use a mulch; vegetables touching the soil may rot.

Clean up crop refuse early.

Plant as early in the spring as practical.

Keep out weeds which harbor insects and diseases.

Summer fallowing (clean cultivation) helps control nematodes.

Summer flooding, where soil type permits, helps control nematodes.

Hand-pick insects.

Water in morning so plants are not wet at night.

Dispose of severely diseased plants before they contaminate others.

Some insects, like cabbage worms, may be killed by spraying with natural preparations such as *Bacillus thuringiensis*.

Rotate Garden Areas

Bake transplanting soil in oven at 160°F for 1 hour. *Crotolaria spectabalis* and marigolds, when planted as cover crops, tend to reduce some kinds of nematodes. The use of marigolds to *repel* nematodes from interplanted vegetables is not effective control. A good garden mulch tends to reduce *damage* caused by nematodes. Many organic gardeners approve of and use sprays and other preparations containing naturally occurring materials. Diatomaceous Earth comes from petrified sea life. Pyrethrin, rotenone, and ryania are examples of natural poisons from plant parts.

These give some control to some insects under certain conditions. Natural predators should be encouraged wherever possible; however, predators raised in captivity, then released into the garden area are usually ineffective. Insecticidal soaps, made from fatty acids tend to work well for some insects under average conditions. Insect traps, baited with phermone lures, work well in some instances. Many of these have sticky adhesives to catch insects. Solar fumigation is effective in reducing some soil-borne problems such as nematodes.

Organic Gardening Supplies

Suitable materials for growing vegetables the organic way are not always easy to locate. The "conventional" garden supply centers carry many products, especially seeds and equipment, which may be used by the organic enthusiast. However, for the difficult-to-find items, the gardener may have to order from specialty businesses dealing in organic gardening supplies.

CROP EVOLUTION AND HUMAN CIVILIZATION

Agriculture evolved independently in many places on this earth, but the earliest evidence of farming dates 10,000 years ago in present day Iraq For much of the 200,000 or so years prior to agriculture, humans lived as nomadic hunters, gatherers, and scavengers surviving solely on wild plants and animals. Subsequent domestication of these wild plants and animals from their natural habitats launched agriculture, thus radically transforming human societies. This occurred initially in the Fertile Crescent, the Andean region in South America, Mexico, and parts of Asia, but diffused throughout much of the globe. A change from the nomadic lifestyle to farming led us to become community dwellers, eventually spawning the development of languages, literature, science, and technology as people were freed from the continuous daily task of finding food.

Some regions caught on much faster than others, by margins of thousands of years Plants have also evolved or, more accurately, they have been changed rapidly by human intervention Every crop plant grown today is related to a wild species occurring naturally in its centre of origin, and progenitors of many of our crops are still found in the wild. Early humans must have tried eating thousands of feral plant species from a pool of a quarter of a million flowering plants before settling down on less than one thousand such species, which were subsequently tamed and adapted to farming.

A little over 100 crop species are now grown intensively around the world, with only a handful of them supplying us with most of what we now eat. Through a process of gradual selection, our ancestors chose a very tiny section of the wild plant community and transformed it into cultivated crops. Some profound alterations in the plant phenotype occurred during such selection, and these include determinate growth habit; elimination of grain shattering; synchronous ripening; shorter maturity; reduction of bitterness and harmful toxins; reduced seed dispersal, sprouting and dormancy; greater productivity, including bigger seed or fruit size; and even an elimination of seeds, such as in banana. These changes reduced the survivability of crops in the wild, and thus a feature that transcends all of our crops is the reduction of weedy traits from wild plants. Present crops are thus totally dependent upon human care for their survival, and modern crop varieties would persist in the wild "no longer than a Chihuahua would last in the company of wolves"

Most crops that supply our food were thus obtained at the end of the Stone Age, often from a relatively narrow pool of extant wild genetic diversity. Additional diversity arose within such cultivated crops through new mutations and natural hybridization, and through judicious selection and perpetuation by farmers who maintained them as land races.Varied uses and preferences brought forth further diversification such as in corn (popcorn, sweet corn, dent corn, broom corn, and flour corn for tortilla and corn bread) or the derivatives of ancestral cabbage (kale, kohl rabi, brussels sprouts, cabbage, cauliflower, and broccoli). With the advent of transoceanic navigation and the "discovery" of the New World, crops were moved around the world rapidly, often achieving prominence in adopted homes far beyond their natural centers of origin or domestication.

For instance, the United States is the leading producer of corn and soybean in the word, yet these crops are native to Mexico and China, respectively. The world's largest traded commodity, coffee, had a humble origin in Ethiopia, but now much of it is produced in Latin America and Asia. Florida oranges have their roots in India, while sugarcane arose in Papua New Guinea. Food crops that are now so integral to the culture or diet in the Old World, such as the potato in Europe, chili pepper in India, cassava in Africa, and sweet potato in Japan, were introduced from South America. For that matter, every crop in North America other than the blueberry, Jerusalem artichoke, sunflower, and squash are borrowed from elsewhere! A few sources of our food are also recent domesticates. Chinese gooseberry occurs wildly in China and is not edible. But careful breeding made it palatable, and it was re-christened "Kiwi fruit" in New Zealand after its introduction there early in the 20th century. The modern strawberry with big fruits is a product of the accidental crossing of two wild species from Virginia (United States) and Chile in France in the mid-18th century.

Rapeseed, grown in India for centuries, was altered recently through classical breeding to eliminate the toxic erucic acid and smelly glucosinolates to result in canola—Canadian oil. Triticale, a completely new crop, was artificially sythesized a few decades ago by combining the genomes of wheat and rye (two distinct genera that do not interbreed in nature). It is now grown on over three million acres worldwide. Modern bread wheat itself is also a fairly recent crop in the evolutionary time scale, having arisen only about 4,000 years ago through hybridization of tetraploid (pasta or durum) wheat with inedible goat grass.

FROM MESOPOTAMIA TO MENDEL

While humans have always molded the evolution of crop plants, such changes imposed by farmers occurred over several millennia, leading to rich crop diversity—especially in traits related to their planting or consumption. At the same time, global population grew very slowly until the mid-19th century. It took 1,800 years for the global population to climb from an estimated 300 million around the time when Christianity began, to reach its first billion. But it took only 12 years to add the last billion, rising from five billion people in 1987 to six billion two years ago. Fortunately, parallel scientific developments in agriculture ensured that food production kept pace with the population explosion of the past century Beginning with Mendel's study of peas, knowledge of genetics helped usher in scientific crop development, resulting in high-yielding varieties. Food production increased in every part of the world in the past few decades, including in Africa. Per capita food consumption has also increased steadily everywhere except in parts of sub-Saharan Africa. In the United States and Canada, where such scientific developments and their applications were most intense, one average farmer now produces enough to feed nearly 150 people! In crops subject to intensive scientific attention—corn, wheat, and rice—the productivity levels increased severalfold. For example, U.S. corn growers averaged 26 bushels of corn per acre in 1928 and 134 bushels per acre in 1998

Such a prodigious increase in agricultural production was underpinned by scientific crop improvement methods along with other developments, including the use of irrigation, improved soil fertility management, mechanization, and control of diseases and pests To develop better crop varieties, scientists have used an array of tools. Artificial crossing, or hybridization, helped us assimilate desirable traits from several varieties into elite cultivars. When desired characteristics were unavailable in the cultivated plants, genes were liberally borrowed from wild relatives and introduced into crop plants. When a crop variety refused to mate with the wild species, various tricks were employed to force them to intermingle, such as the use of the carcinogenic chemical colchicine or by rescuing the hybrid embryos with tissue culture methods. Hybrid vigor was exploited in crops such as corn and cotton to boost productivity. When existing genetic variation within the cultivated germplasm was not adequate, breeders created new variants using ionizing irradiation (gamma ray, x-ray, neutron), mutagenic chemicals (ethyl methane sulfate, mustard gas), or through somaclonal variation

Most people who are concerned about modern biotechnology have little or no knowledge of the processes that have been used to transform crops in the past. Nor are they likely aware that crops have been continually altered over time or that, without human care, they would cease to exist. Using a variety of tools over the past few decades, plant breeders have radically transformed our crop plants by altering their architecture (such as the development of dwarf wheat and rice), shortening growing seasons, developing greater resistance to diseases and pests (all crops), and developing bigger seeds and fruits These crops are also more responsive to management and better adapted to diverse ecological conditions. Improved food quality also resulted through fewer toxins (canola), better digestibility (beans), increased nutrition (high-protein corn), better taste, longer shelf life (thus withstanding long transportation and storage), and enhanced freshness in many vegetables and fruits. A 1,000-fold increase in the marble-sized wild *Lycopersicon* resulted in the modern tomato that can now weigh as much as a kilogram. Modern farming has steadily increased the supply of relatively safe, affordable, and abundant food not only in the developed world, but also in most developing countries.

An average American family now spends only 11% of its income on food and yet has access to better food choices with more variety and nutrition than ever before. Without scientific developments in agriculture, we would otherwise be farming on every square inch of arable land to produce the same amount of food! Using gene transfer techniques to develop GM crops thus can be seen as a logical extension of the continuum of devices we have used to amend our crop plants for millennia. When compared to the gross genetic alterations using wide-species hybridization or the use of mutagenic irradiation, direct introduction of one or a few genes into crops results in subtle and less disruptive changes that are relatively specific and predictable. The process is also clearly more expeditious, as the development of new cultivars by classical breeding typically takes from 10 to 15 years. The primary attraction of the gene transfer methods to the plant breeder, however, is the opportunity to tap into a wide gene pool to borrow traits, obviating the constraints of cross-compatible crop species.

ADDRESSING PUBLIC CONCERNS

While direct gene transfer is still a relatively new approach, many concerns arising from its use may be addressed with the "benchmark" of conventionally bred varieties, as we have the accumulated experience and knowledge with the latter for more than a century. While it seems logical to express a concern such as "I don't know what I am eating with GM foods!" it must be remembered that we really never had that information before with classically bred crops. With GM crops, at least we know what new genetic material is being introduced, so we can test for predictable and even many unpredictable effects. Consider, for example, how conventional plant breeders would develop a disease-resistant tomato. They would introduce chromosome fragments from its wild relative to add a gene for disease resistance. In the process, hundreds of unknown and unwanted genes would also be introduced, with the risk that some of them could encode toxins or allergens, armaments that wild plants deploy to survive. Yet we never routinely tested most conventionally bred varieties for food safety or environmental risk factors, and they were not subject to any regulatory oversight. We have always lived with food risks, but in the last few decades we have become increasingly more adept at asking questions.

To address the concern about long-term health consequences of GM foods, it is instructive to recognize that we worried little about such impacts when massive amounts of new proteins (and unfamiliar chemicals) were introduced into our foods from wild species or when unknown changes were created through mutation breeding. When new foods from exotic crops are introduced, we often assimilate them easily into our diets. What's more we rarely, if ever, before asked the same questions that we now pose about GM crops. Many so-called functional foods, health foods, and nutraceuticals have been entering into the mainstream American diet lately, with little or no regulation or testing. We do not question the long-term health implications of these food supplements, even though they involve relatively large changes in our food intake.

In contrast, the GM foods currently on the market have been tested extensively and judged to be substantially equivalent to their conventional counterparts, with just one or two additional proteins present in miniscule amounts (introduced into a background of thousands of proteins). And, those proteins are broken down either during processing or digestion, with little long-term consequence. In food products such as oil, starch, and sugar, such proteins are not even found. A nagging potential problem with a new protein in food is that it could be a potential allergen. As most food allergens are now well studied, we know that they are found in few defined sources (peanut and other grain legumes, shellfish, tree nuts, and a handful of other foods) and share many similar structural features. Moreover, they must be present in huge proportions in our food, and we must be sensitized to them over time for them to cause any adverse effects. Thus, it is highly unlikely for new allergens to be introduced into our food supply from GM plants.

HISTORICAL ABSENCE

There is no such thing as safe food, and there never has been! That is not to suggest that all of our foods are dangerous, only an acknowledgment that trace levels of such contaminants as toxins and carcinogens are present in everything we eat. But a primary rule of toxicology, articulated over 400 years ago by Paracelsus, refers to the importance of dosage: "Every substance is a poison, but it is the dosage that makes it poisonous" While not alarming, our daily food naturally contains thousands of chemicals, and many of them are shown to be carcinogenic or hazardous in lab animal studies with huge doses. We consume roughly 5,000 to 10,000 natural toxins daily, as plants have evolved to produce an array of chemicals to protect themselves against pests, diseases, and herbivores 7 have been tested and 19 of them found to be rodent carcinogens The fat-soluble neurotoxins solanine and chaconine are present in potatoes and can be detected in the bloodstream of all potato eaters transferring genes through conventional methods, the resistance is often accompanied by an increase in such toxic compounds.

Thus, it is not true that we never had problems with conventionally bred varieties. Any crop variety found to pose a real health risk was promptly removed from the market, but those varieties (in contrast to GM crops) were never routinely tested. One pest-resistant celery variety produced rashes in agricultural workers and subsequently was found to contain 6,200 ppb of carcinogenic psoralens compared to 800 ppb in the control celery This celery was removed from cultivation and that was also the case with the potato variety Lenape, which contained very high levels of toxic solanine. We have always learned from trial and error with all innovations. Similarly, crop improvement practices evolved over time with continued refinement. It is common, though, for human nature to generate an exaggerated fear of new innovations while perceiving older or "natural" products as always more benign. Huber discusses this double standard in the larger context of risk regulation. We have always been lenient towards existing known and greater hazards, even as we create "gatekeepers" to minimize new risks. Thus, we fail to recognize and "exorcise" much larger older risks.

While most food hazards arise from pathogens such as *Escherichia coli* 0:157, *Listeria monocytogenes*, and *Salmonella enterica* along with mycotoxins produced by fungi (and thus a function of food storage and handling), certain foods containing toxic compounds are known to produce adverse health consequences over time. Cassava, eaten by a large population in Africa, contains cyanogenic glucosides, which cause limb paralysis if consumed before extensive processing. Solanin in tomato and potato is known to cause spina bifida. Vetch pea, a common legume known for its hardiness—and thus popular in India among poor farmers—contains highly dangerous neurotoxins that cause untold misery. Phytohemagglutinin, found in undercooked kidney beans, is toxic. And peach seeds are extremely rich in cyanogenic glucosides. None of these were subject to any mandatory testing before they were introduced into the food chain, nor are they subject to any regulation now. But if the current regulatory standards imposed on GM crops were to be invoked for traditional crops, most of them would fail to meet their requirements. Humans have built-in natural defenses that protect us against normal exposure to toxins. But, according to Ames and Gold we have not evolved to achieve "toxic harmony" with everything we eat, because natural selection occurs much too slowly and because much of what is in our diet today was not eaten at all when we were hunter-gatherers. A balanced mixture of foods normally provides adequate nutrition. However, none of the crops grown today were selected with our nutritional requirements in mind. Instead they were chosen intuitively, by our ancestors, from among the edibles that could be found around them. Thus, the most important food crop in the developing world—rice—has no provitamin A and little iron in its endosperm. This has led to horrific problems, such as blindness among millions of children due to vitamin A deficiency, and iron-deficiency anemia in nearly a billion women dependent on a rice diet. Biotechnology research, far from causing any new food safety problems, has already demonstrated its potential in enhancing the nutritional quality of our food and is also being employed to reduce harmful toxic compounds that exist in our food.

ORGANIZED FOOD PRODUCTION

All of us have to eat to live, and organized food production is the most ecologically demanding endeavor we have pursued. Agricultural expansion over the millennia has destroyed millions of acres of forestland around the world. Alien plant species have been introduced into non-native environments to provide food, feed, fiber, and timber, and as a result have disrupted local fauna and flora. Certain aspects of modern farming have had a negative impact on the biodiversity of crop plants and on air, soil, and water quality; nevertheless, it sustains and nurtures most of the world's six billion people with adequate nutrition and affordable food. How can we address the potential environmental concerns of GM crops in the context of our experience with traditional crop variety deployment?

We have continuously introduced genes for disease and pest resistance through conventional breeding into all of our crops. Traits, such as stress tolerance and herbicide resistance, have also been introduced in some crops, and the growth habits of every crop have been altered. The risk of crop gene flow to weedy relatives has always existed, and such "gene flow" occurs where possible. Thus, it is comforting to recognize that no major "superweeds" have developed since the advent of modern plant breeding, although there have been a few instances of crops ever becoming weedy or of weeds becoming more invasive due to gene transfer from crops. Most noxious weeds, such as kudzu, water hyacinth, and parthenium, resulted from the introduction of semidomesticated wild plants into non-native environments without the checks and balances of their native pests. Yet, there are probably no dwarf plants among the wild *Oryza* spp. and *Triticum* spp. populations in the Middle East or Asia, despite the fact that we now have been growing diminutive rice and wheat varieties for decades.

The risk of gene transfer to wild plants is exacerbated when crops are planted in an area with compatible weedy relatives (as often seen in their centers of origin), when such species are promiscuous out-crossers (canola), or, most importantly, when the introduced genes enhance the reproductive fitness of the recipient weeds (although most genes introduced into crop plants, conventional or biotech, have little value in the wild).

The risk of gene transfer to weeds is similar with both conventional and GM crops and is not contingent on how we introduced these genes into plants. We must be vigilant to ensure that weeds do not become noxious as a result of any new crop variety. The current case-by-case testing and monitoring approach with biotech crops is a good regimen for the future, while the past experience with conventional crops provides assurance that such risks will be minimal and manageable. Crop biodiversity is another issue of concern. The popularity of high-yielding varieties has already narrowed the genetic variation found in major crops. Biotechnology, if employed strategically, can reverse this through the recovery of older varieties that were discarded for lack of certain features (such as resistance to new disease strains), because modern gene transfer can restore such traits. Biotechnology research is also enabling the development of better methods for ex situ preservation of germplasm, such as cryopreservation, whereby valuable germplasm is being stored and thus saved from extinction.

The introduction of corn with a single transferred Bt gene has led to some concern about its ecological impact. While this concern should not be dismissed, it should be balanced with our hindsight and experience with corn itself, an introduced alien species now grown on 75 million acres in the United States, where none existed about 1,000 years ago.

A crop introduced into a new environment entails the wholesale introduction of thousands of new genes. When grown on massive amounts of land, it exerts considerable ecological impact on the native fauna and flora, including beneficial insects. In contrast, the introduction of one or two genes into this background of 50,000 genes present in corn will have relatively less effect on the environment.

While the initial fear about the reported damage to monarch butterflies from Bt corn has not held up in additional studies, one also needs to consider the negative impact of alternate practices (such as pesticide sprays) and recognize the potential for positive impacts on beneficial insects by the GM crop due to the specificity of the insect target(s). For that matter, any concern about "gene pollution" pales in comparison to the massive "risk" of alien crop introduction, as 95% of the crop area in the United States now consists of such introduced crops. Concern about horizontal transfer of genes from GM crops to other organisms, such as bacteria, has also been expressed. But it appears highly unlikely that the risk is dependent upon the method of gene introduction. An inherent feature of biotechnology is that it lends itself easily to molecular detection of introduced genes, but a true measure of risk can only come in comparisons with classically bred crops where little or no such studies have been performed.

Concerns such as random gene insertion, gene instability, and genomic disruption due to gene transfer have been expressed, but they are unlikely to be unique to GM crops or of any significance considering our current knowledge of genomic flux in plants. Worries about mixing genes from unrelated species ignore the history of plant breeding and the existing overwhelming sequence similarity of genes across kingdoms. Nevertheless, scientific research aimed at risk analysis, prediction, and prevention, combined with adequate monitoring and stewardship, must continue so that negative ecological impact from GM crops will be kept to a minimum. Most problems raised by science can be solved by additional science itself. For example, appropriate promoters may ensure that pollen will not express genes toxic to beneficial insects, while gene expression strategies, such as sterile pollen, could reduce the risk of gene flow.

IMPACT OF GM CROPS ON THE ENVIRONMENT

One must also recognize the potential positive impact of GM crops on the environment, such as decreasing agricultural expansion to preserve wild ecosystems; improving air, soil, and water quality by promoting reduced tillage, reducing chemical and fuel use; improving biodiversity through resuscitation of older varieties and promotion of beneficial insects; and cleaning up contaminated soil and air through phytoremediation. As we chart ahead with more exciting developments in biotechnology, such as genomics, and grapple with issues arising from consumer acceptance of innovations, historical knowledge on societal adoption of technological innovations may provide some valuable perspectives to scientists. Many innovations that would be good candidates for generating consumer apprehension and concern today were introduced in the past without concern because the public was less informed about innovation. The precautionary principle was never invoked to ensure the scientific certainty that crop varieties developed using nuclear irradiation or chemical mutagens were safe. And food labeling was never demanded for bread wheat improved with the addition of hundreds of unknown goat grass genes.

Many other innovations that are now commonplace in our lives were met with skepticism and opposition when first introduced. Such fear of technology was especially more pronounced in food-related innovations (*e.g.* Pasteurization, canning, freezing, the microwave oven).

However, once consumers recognize that new innovations can enhance their quality of life and once they understood that risks are either minimal or manageable, such technology eventually could enjoy public acceptance. This includes even those "disruptive" technologies that replace older ones (*e.g.* cars versus horse buggies, compact disc versus cassette tape). Nevertheless, there are historical instances of useful innovations that have not been readily accepted due to a variety of reasons, such as recalcitrance to adapt (*e.g.* Dvorak versus QWERTY keyboard), entrenched economic interests opposing change (*e.g.* the metric system in the United States; Beta versus VHS videotape), ideological opposition (*e.g.* plant breeding during Stalin-era Soviet Union by Lysenko), exaggerated notions of risk (*e.g.* food irradiation), ill-timed product introductions, and serious conflicts with societal values and beliefs.

Humans and crops will always be mutually dependent on each other's survival, and the guided evolution of crops will continue but increasingly will be more knowledge-based and responsible. An appreciation of the history of agricultural development however may provide us with a useful roadmap for devising appropriate strategies to informing and rationalizing societal responses to crop improvement. Paraphrasing the American philosopher George Santayana, ignoring history may condemn us to repeat it, but an understanding of the past may as well lead us to an enlightened future.

CROP ALTERATION AND BIOTECHNOLOGY

Crop alteration has been practiced by humankind for thousands of years, since the beginning of civilization. Altering crops through breeding practices changes the genetic make-up of a plant to develop crops with more beneficial characteristics for humans, for example, larger fruits or seeds, drought-tolerance, or resistance to pests. Significant advances in plant breeding ensued after the work of geneticist Gregor Mendel. His work on dominant and recessive alleles gave plant breeders a better understanding of genetics and brought great insights to the techniques utilized by plant breeders.

Crop breeding includes techniques such as plant selection with desirable traits, self-pollination and cross-pollination, and molecular techniques that genetically modify the organism. Domestication of plants has, over the centuries increased yield, improved disease resistance and drought tolerance, eased harvest and improved the taste and nutritional value of crop plants. Careful selection and breeding have had enormous effects on the characteristics of crop plants. Plant selection and breeding in the 1920s and 1930s improved pasture in New Zealand. Extensive X-ray and ultraviolet induced mutagenesis efforts during the 1950s produced the modern commercial varieties of grains such as wheat, corn and barley.The Green Revolution popularized the use of conventional hybridization to increase yield many folds by creating "high-yielding varieties". For example, average yields of corn in the USA have increased from around 2.5 tons per hectare in 1900 to about 9.4 t/ha in 2001. Similarly, worldwide average wheat yields have increased from less than 1 t/ha in 1900 to more than 2.5 t/ha in 1990. South American average wheat yields are around 2 t/ha, African under 1 t/ha, Egypt and Arabia up to 3.5 to 4 t/ha with irrigation. In contrast, the average wheat yield in countries such as France is over 8 t/ha. Variations in yields are due mainly to variation in climate, genetics, and the level of intensive farming techniques.

GENETIC ENGINEERING

Genetically Modified Organisms (GMO) are organisms whose genetic material has been altered by genetic engineering techniques generally known as recombinant DNA technology. Genetic engineering has expanded the genes available to breeders to utilize in creating desired germlines for new crops. After mechanical tomato-harvesters were developed in the early 1960s, agricultural scientists genetically modified tomatoes to be more resistant to mechanical handling. More recently, genetic engineering is being employed in various parts of the world, to create crops with other beneficial traits. New research on woodland strawberry genome was found to be short and easy to manipulate. Researchers now have tools to improve strawberry flavours and aromas of cultivated strawberries.

HERBICIDE-TOLERANT GMO CROPS

Roundup Ready seed has a herbicide resistant gene implanted into its genome that allows the plants to tolerate exposure to glyphosate. Roundup is a trade name for a glyphosate-based product, which is a systemic, nonselective herbicide used to kill weeds. Roundup Ready seeds allow the farmer to grow a crop that can be sprayed with glyphosate to control weeds without harming the resistant crop. Herbicide-tolerant crops are used by farmers worldwide.

Today, 92 per cent of soybean acreage in the US is planted with genetically modified herbicide-tolerant plants. With the increasing use of herbicide-tolerant crops, comes an increase in the use of glyphosate-based herbicide sprays. In some areas glyphosate resistant weeds have developed, causing farmers to switch to other herbicides. Some studies also link widespread glyphosate usage to iron deficiencies in some crops, which is both a crop production and a nutritional quality concern, with potential economic and health implications.

INSECT-RESISTANT GMO CROPS

Other GMO crops used by growers include insect-resistant crops, which have a gene from the soil bacterium *Bacillus thuringiensis* (Bt), which produces a toxin specific to insects. These crops protect plants from damage by insects; one such crop is Starlink. Another is cotton, which accounts for 63 per cent of US cotton acreage. Some believe that similar or better pest-resistance traits can be acquired through traditional breeding practices, and resistance to various pests can be gained through hybridization or cross-pollination with wild species. In some cases, wild species are the primary source of resistance traits; some tomato cultivars that have gained resistance to at least 19 diseases did so through crossing with wild populations of tomatoes.

COSTS AND BENEFITS OF GMOS

Genetic engineers may someday develop transgenic plants which would allow for irrigation, drainage, conservation, sanitary engineering, and maintaining or increasing yields while requiring fewer fossil fuel derived inputs than conventional crops. Such developments would be particularly important in areas which are normally arid and rely upon constant irrigation, and on large scale farms.

However, genetic engineering of plants has proven to be controversial. Many issues surrounding food security and environmental impacts have risen regarding GMO practices. For example, GMOs are questioned by some ecologists and economists concerned with GMO practices such as terminator seeds, which is a genetic modification that creates sterile seeds.

Terminator seeds are currently under strong international opposition and face continual efforts of global bans. Another controversial issue is the patent protection given to companies that develop new types of seed using genetic engineering. Since companies have intellectual ownership of their seeds, they have the power to dictate terms and conditions of their patented product. Currently, ten seed companies control over two-thirds of the global seed sales.

Vandana Shiva argues that these companies are guilty of biopiracy by patenting life and exploiting organisms for profit. Farmers using patented seed are restricted from saving seed for subsequent plantings, which forces farmers to buy new seed every year. Since seed saving is a traditional practice for many farmers in both developing and developed countries, GMO seeds legally bind farmers to change their seed saving practices to buying new seed every year. Locally adapted seeds are an essential heritage that has the potential to be lost with current hybridized crops and GMOs.

Locally adapted seeds, also called land races or crop eco-types, are important because they have adapted over time to the specific micro-climates, soils, other environmental conditions, field designs, and ethnic preference indigenous to the exact area of cultivation.

Introducing GMOs and hybridized commercial seed to an area brings the risk of cross-pollination with local land races. Therefore, GMOs pose a threat to the sustainability of land races and the ethnic heritage of cultures. Once seed contains transgenic material, it becomes subject to the conditions of the seed company that owns the patent of the transgenic material.

SQUARE FOOT GARDENING

Square Foot Gardening is the practice of planning small but intensively planted gardens. The phrase "square foot gardening" was popularized by Mel Bartholomew. The practice combines concepts from other organic gardening methods, including a strong focus on compost, closely planted raised beds and biointensive attention to a small, clearly defined area. Proponents claim that the method is particularly well-suited for areas with poor soil, beginning gardeners or as adaptive recreation for those with disabilities.

BASIC METHOD

The original square-foot-gardening method used an open-bottomed box to contain a finite amount of soil, which was divided with a grid into sections. To encourage variety of different crops over time, each square would be planted with a different kind of plant, the number of plants per square depending on an individual plant's size. A single tomato plant might take a full square, as might herbs such as oregano, basil or mint, while most strawberry plants could be planted four per square, with up to sixteen radishes per square. Tall or climbing plants such as maize or pole beans might be planted in a northern row so as not to shade other plants, and supported with lattice or netting.

The logic behind using smaller beds is that they are easily adapted, and the gardener can easily reach the entire area, without stepping on and compacting the soil. In the second edition, Bartholomew suggests using a "weed barrier" beneath the box, and filling it completely with "Mel's mix," a combination by volume of one third of decayed Sphagnum "peat moss", one-third expanded vermiculite and one-third blended compost. For accessibility, raised boxes may have bottoms to sit like tables at a convenient height, with approximately 6" of manufactured soil per square foot.

RATIONALE

In this method, the garden space is divided into beds that are easily accessed from every side. A 4' x 4', 16 sq ft or 120 cm x 120 cm, 1.4 m^2 garden is recommended for the first garden, and a path wide enough to comfortably work from should be made on each side of the bed, if possible, or if the bed must be accessed by reaching across it, a narrower one should be used so that no discomfort results from tending the garden. Each of the beds is divided into approximately one square foot units and marked out with sticks, twine, or sturdy slats to ensure that the square foot units remain visible as the garden matures.

Different seeds are planted in each square, to ensure a rational amount of each type of crop is grown, and to conserve seeds instead of overplanting, crowding and thinning plants. Common spacing is one plant per square for larger plants, four plants per square for medium large plants like lettuce, nine plants per square for medium-small plants like spinach, and sixteen per square for small plants such as onions and carrots.

Plants that normally take up yards of space as runners, such as squash or cucumbers, are grown vertically on sturdy frames that are hung with netting or string to support the developing crops. Ones that grow deep underground, such as potatoes or carrots, are grown in a square foot section that has foot tall sides and a planting surface above the ground, so that a foot or more of framed soil depth is provided above the garden surface rather than below it.

The beds are weeded and watered from the pathways, so the garden soil is never stepped on or compacted. Because a new soil mixture is used to create the garden, and a few handfuls of compost are added with each harvest to maintain soil fertility over time, the state of the site's underlying soil is irrelevant. This gardening method has been employed successfully in every region, including in deserts, on high arid mountain plateaus, in cramped urban locations, and in areas with polluted or high salinity soils. It is equally useful for growing flowers, vegetables, herbs and some fruits in containers, raised beds, on tabletops or at ground level, in only 4 to 6 inches of soil. A few seeds per square foot, the ability to make compost, to water by hand, and to set up the initial garden in a sunny position or where a container, table or platform garden may be moved on wheels to receive light is all that is needed to set up a square foot garden.

BENEFITS OF SQUARE FOOT GARDENING

- *Much less work*: Conventional gardening requires heavy tools to loosen the soil, whereas in this method, the soil is never compacted and it remains loose and loamy. Weeding takes only seconds to minutes, due to the light soil, raised beds, and easily accessed plants. Harvests per foot of garden are increased due to the rich soil mixture, well-spaced plants, and lack of weeds produced when following Mel Bartholomew's method.
- *Water Savings*: The soil mixture that is advised has water-holding capacities, so that the garden needs water less frequently, and in much smaller quantities than when using other gardening methods. Water is also spared by hand-watering directly at the plant roots, so that there is very little waste and tender young plants and seedlings are preserved.
- *Very little weeding*: One benefit of this close planting is that the vegetables form a living mulch, and shade out many weed seeds before they have a chance to germinate.
- *Pesticide/ Herbicide Free*: Natural insect repellent methods like companion planting become very efficient in a close space and thus, pesticides are not necessary. The large variety of crops in a small space also prevents plant diseases from spreading easily.
- *Accessibility*: A plywood bottom can be attached to the bottom of a box, which can then be placed on a tabletop or raised platform for those who wish to garden without bending or squatting, or to make gardening easy for wheelchair, cane or walker users.

SUCCESSION PLANTING

Succession planting is done so garden space is put to its most efficient use. With this method, there is never an unplanted area in the vegetable plot. As soon as one crop is harvested, the space is immediately replanted with a different crop. This method of succession planting uses groupings of cool-season and warm-season crops. In the simplest plan, succession planting involves planning for spring, summer and fall crops.

Another method is to make several plantings of a vegetable. Bush beans can be planted every 2 weeks from mid-May to the beginning of August. A third type of succession planting involves a single planting of several different cultivars with varying maturity dates. This method works well for cabbage and corn. These methods allow for a continuous harvest over a longer period.

Vegetable harvest can be extended into the fall by growing the faster growing, cool-season crops. The quality of many of the cool-season crops is exceptional when they mature in the fall. Plant so there is adequate time for the crop to mature. When planting seeds for a fall crop, you must water the seeds more often than in the spring because the summer soil is usually warm and dry. Provide some shade to help retain moisture. Cover the seedbed with cardboard or newspaper. Check the seedbed daily and remove the covering when the first seed sprouts. Keep the bed consistently watered until all the seedlings are established. Pre-sprouting also ensures good germination of cool-season crops in the hot, dry soil of August.

Bibliography

A. Bandyopadhyay , K.V. Sundaram , M. Moni , M.M. Jha and P.S. Kundu: *Sustainable Agriculture*, Northern Book Centre, Delhi, 2003.

A. Rashid: *Introduction to Genetic Engineering of Crop Plants : Aims and Achievements*, I.K. International Publication, Delhi, 2009.

B K Desai and B T Pujari: *Sustainable Agriculture : A Vision for Future*, New India Publication, Delhi, 2007.

C. Rajendran, K. Ramamoorthy and S. Juliet Hepziba: *Nutritional and Physiological Disorders in Crop Plants*, Scientific Publication, Delhi, 2009.

Franklin P. Gardner, R. Brent Pearce and Roger L. Mitchell: *Physiology of Crop Plants*, Scientific Publication, Delhi, 2010.

G. Kalloo, Mathura Rai, Major Singh and Sanjeet Kumar: *Heterosis in Crop Plants*, Researchco Book Center, 2006.

Herbert Kendall Hayes and Ralph John Garber: *Breeding Crop Plants*, Asiatic Publication, Delhi, 2008.

K Dasarathararamaiah and M Jayaraj: *Sustainable Agriculture in Drought Prone Regions : Papers in Honour of Prof. N. Narayana*, Serials Publication, Delhi, 2007.

Madan Pal Singh, Sangeeta Khetarpal, Renu Pandey and Pramod Kumar: *Climate Change : Impacts and Adaptations in Crop Plants*, Today and Tomorrow Publication, Delhi, 2011.

P N Kalla; Anita Singh; S S Pareek; Shanti K Sharma and Hanuman Ram: *Sustainable Agriculture : Status and Prospects*, Agrotech Publishing Academy, 2007.

P. Sudhakar, P. Latha and P.V. Reddy: *Phenotyping Crop Plants for Physiological and Biochemical Traits*, BS Publications, Delhi, 2014.

Pratap Singh and P.L. Maliwal: *Technologies for Food Security and Sustainable Agriculture*, Agrotech Publication, Delhi, 2005.

Pravin Chandra Trivedi: *Plant Pest Management*, Aavishkar Publication, Delhi, 2002.

R. Arora, B. Singh and A.K. Dhawan: *Theory and Practice of Integrated Pest Management*, Scientific Publication, Delhi, 2012.

R. Arunachalam and R. Netaji Seetharaman: *Sustainable Agriculture : Indigenous Practice for Natural Resource Management*, Agrobios Publication, Delhi, 2004.

R.A. Raju: *Synthetic Fertilizers for Sustainable Agriculture*, Agrobios Publication, Delhi, 2005.

R.K. Behl, D.P. Singh, V.S. Tomar, M.S. Bhale, D. Khare and S.D. Upadhyaya: *Sustainable Agriculture for Food, Bio-Energy and Livelihood Security*, Agrobios Publication, Delhi, 2009.

Rajiv K. Sinha: *Sustainable Agriculture : Embarking on the Second Green Revolution with Technological Revival of Traditional Agriculture*, Surabhi Publications, Delhi, 2004.

Ramesh Umrani and C.K. Jain: *Sustainable Agriculture*, Oxford Book Company, Delhi, 2010.

S Ignacimuthu & Alok Sen: *Strategies in Integrated Pest Management : Current Trends and Future Prospects*, Phoenix, 2002.

S. Ignacimuthu, s.j. and S. Jayaraj: *Recent Trends in Insect Pest Management*, Elite Publication, Delhi, 2008.

S.M. Paul Khurana: *Pathological Problems of Economic Crop Plants and Their Management*, Scientific Publication, Delhi, 1998.

Shubhrata R. Mishra: *Plant Protection and Pest Management*, Discovery Publishing House, Delhi, 2005.

Sunday Ekesi and Nguya K. Maniania: *Use of Entomopathogenic Fungi in Biological Pest Management*, Research Signpost, Delhi, 2007.

T.V. Sathe and A.D. Jadhav: *Sericulture and Pest Management*, Daya Publication, Delhi, 2001.

Uma Shankar; Satya Priya and Deepak Kumar: *Vegetable Pest Management Guide for Farmers*, International Book, Delhi, 2008.

Vandana Shiva and Gitanjali Bedi: *Sustainable Agriculture and Food Security : The Impact of Globalisation*, Sage Publication, Delhi, 2002.

Vir Singh, Gaurav Papnai and Govind Singh Kushwaha: *Sustainable Agriculture: A Search for Innovations*, Biotech Books, Delhi, 2014.

Index

P

Q

R

S

T

V

W